The Artful Universe

E X P A N D E D

The Artful Universe

EXPANDED

JOHN D. BARROW

Centre for Mathematical Sciences
University of Cambridge

OXFORD
UNIVERSITY PRESS

OXFORD

UNIVERSITY PRESS

Great Clarendon Street, Oxford OX2 6DP

Oxford University Press is a department of the University of Oxford.
It furthers the University's objective of excellence in research, scholarship,
and education by publishing worldwide in

Oxford New York

Auckland Cape Town Dar es Salaam Hong Kong Karachi Kuala Lumpur
Madrid Melbourne Mexico City Nairobi New Delhi Shanghai Taipei Toronto

With offices in

Argentina Austria Brazil Chile Czech Republic France Greece
Guatemala Hungary Italy Japan South Korea Poland Portugal
Singapore Switzerland Thailand Turkey Ukraine Vietnam

Oxford is a registered trade mark of Oxford University Press
in the UK and in certain other countries

Published in the United States
by Oxford University Press Inc., New York

British Library Cataloguing in Publication Data

Data available

Library of Congress Cataloging in Publication Data

Data available

ISBN 0–19–280569–X

1

Typeset by RefineCatch Limited, Bungay, Suffolk
Printed in Great Britain by
Clays Ltd., St Ives plc

To WJT, with thanks

T. S. Eliot was once climbing into a London taxi when the driver said, 'You're T. S. Eliot.' The astonished poet asked how he knew. 'Oh', replied the driver, 'I've got an eye for a celebrity. Only the other evening I picked up Bertrand Russell, and I said to him: "Well, Lord Russell, what's it all about?" and, do you know, *he couldn't tell me.*'

JOHN NAUGHTON

Preface

Life, like science and art, is a theory about the world: a theory that in our case takes bodily form. By a succession of adaptations, most of which are favourable and none of which are lethal, living things have invested in particular expectations about the future course of their environments. If those theories are good enough, then life will prosper and multiply; but if they are outmoded by changing conditions, their embodiments will dwindle and perish.

Science and art are two things most uniquely human. They witness to a desire to see beyond the seen. They display the crowning successes of the objective and subjective views of the world. But while they spring from a shared source—the careful observation of things—they evoke different theories about the world: what it means, what its inner connections truly are, and what we should judge as important.

Science and art have diverged. As science became more successful in its quest to explain the seen by unseen laws of Nature, so art became increasingly subjective, metaphorical, and divorced from realistic representation. It explored other worlds, leaving science to deal with this one. But there is more to art appreciation than the appreciation of art. The sciences can illuminate our penchant for artistic creation. Conversely, the growing fascination of scientists with the fruits of organized complexity in all its forms should draw them towards the creative arts where there are extraordinary examples of structured intricacy. This book is an attempt to look with a scientist's eye at a few things that are usually kept out of scientific view. Things that are admired rather than explained.

Environmentalism is the flavour of the month. Accordingly, we need to appreciate how the cosmic environment imprints itself upon our minds and bodies in ways that shape their structures, their fascinations, and their biases. Astronomers have revealed that we live in a Universe that is big and old, dark and cold; yet it could be no other way. For we shall find that these stark facts of cosmic life are essential if the Universe is to harbour life at all. And from these life-supporting features flows a particular perception of the Universe that we may well share with all its perceivers, whoever they may be. We shall delve into some of the ways in which the structure of the Universe influences the tenor of our philosophizing and feeling for the Universe; what the unsuspected metaphysical impact of the discovery of extraterrestrial life might prove to be; how the inevitable features of a life-supporting planet filter down to influence the structure and behaviour of living things; and how the stars and the sky, overlain by our interpretations of them, have influenced our concepts of time and determinism. These investigations will take us on down unexpected byways to consider how our past environment has fashioned concepts of favourable environments which, in turn, influence our artistic appreciation of landscape. This will reveal new things about our ambiguous attraction to works of computer art and lead us to explore an ancient analogy of the problem of whether computer art is truly art. We will also see why natural colours originate, and how they have helped fashion the colour vision of living things and influenced the symbolic use of colour in modern art and society. Turning from sight to sound, we will consider the origins of music. Music has the power to influence human emotions in ways that other forms of organized complexity do not. In our explorations of its sources and structure, we shall find tantalizing evidence for a common factor behind all humanly enjoyable music that links it, and us, to the overall structure of the environment.

Anthropologists and social scientists have traditionally laid great stress upon the diversity of human artistic and social activity, but largely ignored the common features of existence that derive from the universality of our cosmic environment, and the necessary features that life-supporting environments must display. Just as science has for too long focused almost exclusively upon the regularities and simplicities of the world at the expense of the irregularities and complexities, so our contemplation of the arts has over-indulged the diversities and unpredictabilities of its forms at the expense of the skein of shared features that bind us with these forms of complexity to the underlying environment that the Universe provides. The study of human actions, human minds, and human creativity has been quick to see complexity, slow to appreciate simplicity. Science, brought up to reflect on symmetry, has at long last begun to appreciate diversity. In the fruits of creative activity, science will find the most impressive examples of organized complexity, whilst offering, in return, a new perspective on the sources of our senses, our tastes, and the sights and sounds that surround them.

Many people have helped this project, directly or indirectly, knowingly or unknowingly, at different stages. I would like to thank Mark Bailey, Margaret Boden, Laura Brown, Guiseppe Caglioti, Paul Davies, John Grandidge, Mike Land, John Manger, the late John Maynard Smith, Sir William McCrea, Stephen Medcalf, Jim Message, Leon Mestel, Geoff Miller, Marjorie Mueller, Andrew Murray, Carl Murray, Keith O'Nions, Mike O'Shea, Tim Roper, Robert Smith, David Streeter, Debbie Sutcliffe, the late Roger Tayler, Frank Tipler, and Tatyana Tchuvilyova.

Family members are always puzzled by writers, since they appear to be people for whom writing is harder than it is for others. My wife Elizabeth helped in innumerable patient ways; our children, David, Roger, and Louise, watched with interest, expressed scepticism as to how anyone could call themselves a scientist yet not play computer games or operate the video recorder, and announced that they will soon be writing their own books anyway.

Brighton   J.D.B.
April 1995

Preface to new edition

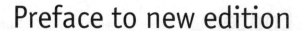

I was pleased that Oxford University Press were keen to publish a new edition of *The Artful Universe* to meet continual demand for the original. Since it first appeared in 1995 there have been a number of interesting developments in areas of science and art that were touched upon in the first edition. I have taken the opportunity to extend the original book in many places to take into account these developments. I am particularly grateful to Michael Rodgers, whose idea it was to begin, to Marsha Filion, whose idea it was to finish, to Latha Menon whose task it was to finish, and to Richard Taylor, Paul Davies, Janna Levin, Nick Mee, Richard Bright, Jean-Pierre Boon, Geoff West, Jayanth Banavar, and Martin Kemp for the discussions and contributions that helped create what happened in between.

John D. Barrow, *Cambridge*, August 2004

Contents

1 | Tales of the unexpected

Arguments against new ideas generally pass through three distinct stages, from 'It's not true', to 'Well, it may be true, but it's not important', to 'It's true and it's important, but it's not new—we knew it all along.'

UNPOPULAR WISDOM

We are inveterate spectators. Large fractions of our lives are spent watching people acting, competing, working, performing, or just simply relaxing. Nor is our interest confined to the human spectacle. We are captivated by 'things' as well: pictures, sculptures, photographs of past experiences—all have the power to capture our attention. And, if we can't watch real life, then we are drawn into the virtual worlds of the cinema, television pictures, and videos. You may even find yourself reading a book.

While some people are skilled in the creation of interesting sights and sounds, others are trained observers. They seek out unusual sights, or register events that most of us would never notice. Some, with the help of artificial sensors, delve deeper and range farther than our unaided senses allow. Out of these sensations has emerged an embroidery of artistic activities that are uniquely human. But, paradoxically, from the same source has flowed a systematic study of Nature that we call science. Their common origins may seem surprising to many, because a great gulf seems to lie between them, shored up by our educational systems and prejudices. The sciences paint an impersonal and objective account of the world, deliberately devoid of 'meaning', telling us about the origins and mechanics of life, by revealing nothing of the joys and sorrows of living. By contrast, the creative arts encode the antithesis of the scientific world-view: an untrammelled celebration of that human subjectivity that divides us from the beasts; a unique expression of the human mind that sets it apart from the unfeeling whirl of electrons and galaxies that scientists assure us is the way of the world.

This book is an attempt to see things differently. We want to explore some of the ways in which our common experience of living in the Universe rubs off on us. It has become fashionable since the 1960s to regard all interesting human attributes as things that are learned from our contacts with individuals and society—as the results of nurture, not nature—and to ignore the universalities of human thinking. Recently, this prejudice has been seriously undermined. Things are far more complicated. The complexities of our minds and bodies witness to a long history of subtle adaptations to the nature of the world and its other occupants. Human beings, together with all their likes and dislikes, their senses and sensibilities, did not fall ready-made from the sky; nor were they born with minds and bodies that bear no imprints of the history of their species. Many of our abilities and susceptibilities are specific adaptations to ancient environmental problems, rather than separate manifestations of a general intelligence for all seasons.

We have instincts and propensities that bear subtle testimony to the universalities of our own environment, and that of our distant ancestors. Some of these instincts, like that for language, are so important that, for all practical purposes, they are unalterable; others are more malleable, and can be partially overwritten, or totally reprogrammed, by experience: they appear as defaults only when cultural influences, or other learned responses, are absent. Some of those environmental universalities stretch farther than our home planet. They reflect the regularities of solar systems, galaxies, and whole universes. They may tell us important things about any form of living intelligence—wherever it might be in the Universe.

To unravel all these strands is an impossible task. Our aims in this book are more modest. We are going to look at some of the unexpected ways in which the structure of the Universe—its laws, its environments, its astronomical appearance—imprints itself upon our thoughts, our aesthetic preferences, and our views about the nature of things. In some instances, those cosmic influences will fix the environments of living things in inevitable ways; in others, our propensities will arise as by-products of adaptations to situations that no longer challenge us. Those adaptations remain with us, often in transmogrified forms, as living evidence of the presence of the past.

There has always been a divide between those who view science as the discovery of real things and those who regard it as an elaborate mental creation designed to make sense of some unknowable reality. The former view is attractive to the scientists because it makes them feel good about what they are doing: exploring unknown territory and unearthing new facts about reality. The latter viewpoint is more readily adopted by those involved in the study of human behaviour. Sociologists and psychologists are so impressed by the inventiveness of the human mind, and by the collective human activities of scientists, that they think that this is all there is to it. But while science certainly embodies those human elements, it is an unjustified leap of logic to conclude that it is therefore nothing but those human elements. This emphasis upon science as just another human activity, rather than a process that involves discovery, can be a subtle manifestation of opposition to the scientific enterprise by downgrading the status of what it does. After all, landscape gardeners don't seem quite as exciting as explorers.

Whatever the strengths of these claims and counterclaims, there is undoubtedly a dilemma for the outsider. Are the sciences and the humanities alternative responses to the world in which we live? Are they irreconcilable? Must we embrace either the subjective or the objective: the abacus or the rose? Or have we created a false dichotomy and are the two views of the world more intimately entwined than appears at first sight?

One of our goals here is to illuminate the relationship between the sciences and the arts with a new perspective on our emergence in the Universe. The fact that we have evolved in a particular type of universe constrains what we think, and how we think, in unsuspected ways. What games and puzzles do we find challenging? Why do we like certain types of art or music? Why do we have a propensity for seeing patterns where none exist? Why do so many myths and legends have common factors? How are these things influenced by our experience of time and space, and by the appearance of the heavens? What is the influence of our characteristic life-span—neither very short, nor very long—upon our thinking about the world, and the value we place upon life? How does the structure of our minds determine what philosophical problems we find challenging? Why are some images so attractive to the human eye? How have the concepts of chance and randomness influenced our religious and ethical thinking? What are the sources of fatalism and our views about the end of the world? If we were to make contact with extraterrestrial civilizations, what might we expect them to like, and be like? What could we learn about them from their aesthetics? Whereas most people foresee great scientific advances flowing from contacts with advanced extraterrestrials, we shall discover that the greatest gains might turn out to be quite different. It is also tempting to adopt a variety of cosmic ageism, which has great expectations about long-lived extraterrestrials. Given world enough and time, we confidently expect them to get closer and closer to uncovering all there is to know about what makes the Universe tick. This optimism may be displaced. If you want to understand the Universe, intelligence and longevity may not be enough. Our own scientific development will be seen to hinge upon a number of extraordinary coincidences about our environment and our view of the sky. In the absence of those fortuitous circumstances, our understanding of the world would be greatly diminished, and our beliefs about the meaning of our own place in the scheme of things radically altered. Moreover, there is evidence to suggest that a certain degree of irrationality may be more than an embarrassing by-product of the evolution of intelligence: it may be an essential feature of progress in natural environments.

In our quest to unpick the Universe's influences upon us, we have far to go. We shall begin by looking at the matter of perspective—our ways of looking at the world. The importance of the vantage-point of the spectator was recognized in art before it was even raised in science. Scientists liked to see themselves as bird-watchers cosseted in a perfect hide. When confronted by the impact of their perceiving upon what was perceived, the certainty of their interpretations of the world was sorely challenged. In retrospect, the ground of our confidence in the reliability of our view of a significant area of the world was established by the discovery that living things evolve and adapt to their environments. We are accustomed to think of environments as local and immediate. Here, we shall discover how our existence derives from a cosmic environment that is billions of light-years in size. If life is to be possible within them, universes must have particular forms. When conscious life does emerge, its experiences and conceptions are strangely influenced by the fact that the Universe must be big and old, dark and cold.

Our next exploration will be into the sizes of things. We shall discover something of the network of interrelationships between living things and the necessary aspects of environments that are life-supporting. This path will take us back in time to the origins of humanity; but, at the end of this road, we shall find unexpected clues about the origins of aesthetics, the haunting appeal of pictures and landscapes, and the importance of symmetry for living things. These insights will shed new light on our responses to modern computer-generated art, and will help us to appreciate what we require of man-made landscapes if they are to soothe or stimulate us.

Our third excursion takes us to the stars: to unveil the ways in which the heavenly clockwork has influenced the nurture and nature of life on Earth. Living things respond to a symphony of celestial rhythms. Over millions of years they have internalized many of those rhythms. With the coming of consciousness and culture, they have responded differently, but no less impressively, to their beat. From the things that are not seen, we move to the things that are seen. The appearance of the night sky is a universal experience. Some of its influences are direct and unnoticed; others are conjured up by our own imaginations. These nocturnal imaginings depend, in crucial ways, upon where—and when—you live. And when the night departs, the day comes, bringing light—coloured light. Light and life combine in ways that enable us to understand our perceptions of colour and some of its deep psychological influences upon us.

In surprising ways, our systems of timekeeping also mask ancient astrological leanings, which have resisted all efforts of principalities and powers to redefine them. Tabloid newspapers and popular magazines still perpetuate the myths of astrology. Ironically, we shall find that, while the constellations can tell us nothing about the future, they have much to tell us about the past.

We then turn from sight to sound, to explore the origins of our susceptibility to music. Why do we like it? Where did it come from? Full of sound and fury, does it signify anything? These are some of the questions that guide our search for an understanding of what music is, and in what ways its universal appeal might be an inevitable by-product of adaptations to other aspects of the environment around us.

The humanities are not manifestations of human creativity alone. Aesthetics and cultural development can find themselves constrained by a mind-cage imposed by our physical nature and by the universality of the cosmic environment in which we have our being. The arts and the sciences flow from a single source; they are informed by the same reality; and their insights are linked in ways that make them look less and less like alternatives.

2 | The impact of evolution

Painting is the art of protecting flat surfaces from the weather and exposing them to the critics.

AMBROSE BIERCE

A room with a view: matters of perspective

The headline in the *Parrot's Weekly* read: *Titanic* Sunk, No Parrots Hurt.

Katherine Whitehorn

Imagination—the making of images—lies at the root of all human creativity, and directs our conscious experience of the world. From early childhood, we are constantly making pictures of things, of people, and of places. As we grow older, we learn new ways of doing it. Photography, painting, descriptive writing, sculpture, poetry: all are means of capturing images in permanent form, so that we can savour and re-experience the fruits of our imagination. But the creative arts are not the only manifestations of the imaginative urge. Science is another quest to make images of the world. It has different goals, and often requires different skills, but its beginnings had much in common with those of art: the accurate observation and representation of the world. Yet, there is more to the world than meets the eye. The accuracy of our perceptions of the world is not something that we can take for granted. Illusion is the dark side of imagination, and illusion tempts with self-delusion, under whose command we cannot long survive. The use of imagination to enlarge our picture of reality without, at the same time, subverting it is a delicate enterprise.

As soon as we start questioning the reliability of our impressions of the world—asking whether the politician or the car salesman is really all he appears to be, or whether the hot desert road is really leading to a vast oasis—we have become philosophers. For centuries philosophers have agonized over whether we can have confidence in our images of the world. In so doing, they have often taken too little heed of *why* we have a view of the world, and from whence it came. Our minds have not fallen ready-made from the sky. They have a history that weds them to the nature of the environment in deep and influential ways.

By uncovering some of the purposes for which our minds developed, and the extent of the environment to which they must adapt, we can shed new light upon the thoughts that minds can have. We shall find that our 'environment' extends farther and wider than we might ever have suspected—impressing its nature upon the direction of our thought, shaping our views about ourselves and the Universe in which we live.

Appraising the world is a matter of perspective. Look at an ancient Egyptian painting (Plate 1) and it looks distinctly odd: awkward and unrealistic, as if someone has squashed the scene flat against the wall. Part of the charm of pictures produced by very young children is the *naïveté* of this same depthless appearance (see Figure 2.1). What these drawings lack is a sense of perspective: the presentation of three-dimensional spatial information on a flat surface. Our eyes are immediately sensitive to its absence or imperfect presence: it is the touchstone of realism in representational art. Traditionally, the systematic use of perspective is traced from its display in a work by Masaccio, painted between 1424 and 1426, called *The Rendering of the Tribute Money* (Plate 2).

Here, three separate scenes are given relative depth by the device of creating a distant point (the 'vanishing-point') to which all lines of sight appear to converge. The effect is enhanced by reducing the intensity of colours in the background.

2.1 An absence of perspective characterizes images made by young children. This picture is by Danny Palmer, age 9.

Although Masaccio died while still in his early twenties, his systematic construction of a realistic perspective challenged others to create accurate representations of objects in three-dimensional space. Piero della Francesca drew his inspiration from Filippo Brunelleschi's studies in architectural perspective and Masaccio's work; he perfected the artistic organization of space by combining lines parallel to the sides of the picture with lines directed towards the vanishing-point. The viewer feels that he is looking in upon the world through an open window (Plate 3).

Renaissance artists developed the geometrical intuitions that are required to create a three-dimensional perspective on a two-dimensional surface, and joined sculptors in bringing the observer into a closer relationship with the things portrayed. But that relationship was still one of separation. The creation of perspective removes the viewer from the scene portrayed within the frame; with that comes an inevitable subjectivity. We are left outside, looking in. This separation of the scene from the observer had parallels in more abstract contemplations of the relationship between the human mind and the outside world. European philosophers, beginning with Descartes, maintained a clear division between the observer and the observed. Our perception of the world cast us in the role of perfectly concealed viewers. No observation of the world could alter its character: the outside world really was outside. But not all cultures reflected this separation of the perceiver from the perceived. Chinese landscape painting manifests an engaging approach to the relationship between three-dimensional space and its representation in two dimensions. It did not introduce linear perspective in the form found in the West, in which the observer's viewpoint is located at a point outside the picture, in front of the canvas. Instead, it lies ambiguously within the landscape. One cannot tell where the observer is situated in relation to the mountains and streams portrayed. Thus, one becomes part of the scene, just as the artist felt himself to be at one with what he was representing. Chinese landscapes deliberately leave the observer bereft of clues as to his location in the picture. We must study the whole picture if the mind is to find its vantage point. The search for the elusive perspective encourages many different readings of the picture and defies attempts to endow it with a single message (Figure 2.2).

Another form of visual subtlety in these oriental landscapes is the absence of shadow. Shadow enhances the illusion of perspective by endowing the observer with a privileged position in space or time, determined by the length and direction of the shadows cast by the Sun's rays. The contrast between a shadowless oriental work and a Western master of the use of shadow, like Rembrandt or Vermeer, could not be greater.

The drawing together of the observer and the observed into a contemplative nexus, mediated by an ambiguity of perspective, reflects the tenor of much

2.2 Chiang Yee's *Cows in Derwentwater* from *The Silent Traveller, a Chinese Artist in Lakeland,* first published in London in 1937.

Eastern art. It seeks to enhance our mediation of natural beauty, rather than merely celebrate our power to replicate it in another static medium. This emphasis upon the act of observation is striking. Whereas a work of Western art would be displayed continuously, a delicate oriental silkscreen might be unrolled only for occasional periods of silent solitary meditation.

The most extreme violation of the Western separation of the medium from the message is to be found in an art-form like Japanese origami. Whereas Western art focuses upon the freedom to move images around on paper or canvas to create fixed patterns, origami ignores that separation between the image and the paper. The paper becomes part of the image, and is twisted and folded until it *is* the picture, not merely the surface on which it lies.

Another deep difference in the Eastern and Western attitudes towards the observer and the observed can be seen in the spontaneity required of the artist. In the West, the development of oil-based paints allowed the artist to evolve and revise his work over a long period of time. He was no longer captive to the irrevocable nature of the medium like the fresco painter or the water-colourist. But such unceasing revisionism was not an acceptable response for the oriental artist. Exquisite Japanese sumi-e ink work was executed in a single uninter-rupted stroke of the brush on the paper, capturing the thought of the moment in an instant irredeemable flourish. The sense of time and development is to be found, not in the revisions and re-enactments of the artist, constantly refining his picture of the world, but in representations of natural change.

The art of bonsai represents this temporal aspect by entraining the natural growth of things through a skilful horticultural intervention. In miniature, it symbolizes the living, growing—but unfinished—nature of the world. It stands in stark contrast to the emphasis of many early Western art-forms. There, the eye was invited to contemplate the completeness and perfection in the arrangement of things, whether in an idyllic landscape or a matrix of religious symbols.

Another contrast between early Western and Eastern representations of the world is highlighted by a trend that developed in Europe over many centuries following the Renaissance. Whereas medieval art had been heavily symbolic in its religious messages, and oriental art placed heavy stress upon the use of delicate compositional harmony as an aid to meditation, in later Western art a quest for realism began. Instead of organizing symbols on a canvas to impart a message that only those versed in the symbolism could decode, Western artists set their sights upon the perfection of their representation of the image that the eye had recorded. This involves two vital skills, which are all the more challenging to acquire because they are diametrically opposed. On the one hand, realism requires an advanced knowledge of geometry, perspective, and the behaviour of light. But on the other, it requires us to empty ourselves of our understanding of *what* is being represented. If we believe the child we are drawing to be divine, then this will influence our representation in ways that obscure the aims of literal realism. From the sixteenth century to the middle of the nineteenth century, Western artists converged upon methods that produced increasingly realistic works by the refinement of subtle techniques of shading and perspective. So influential did this work become, that it set standards for realism by which all subsequent works have been judged, and led us to regard realism as the pinnacle to which all previous techniques were ascending. Yet despite its familiarity, realism is something of a sophisticated novelty that did not develop in cultures that lacked sophisticated geometrical and optical knowledge. This emphasizes the gulf that lies between the process of seeing the world clearly and accurately (which most of us believe we do), and producing an accurate drawing of what we perceive. We lose sight of the real image and add all sorts of changes and corrections to the message our eyes are trying to give us. If we look at some very early art-forms, we get the impression that the idea of trying to match the image with reality never came into play and all that remained as the final rendering of the things seen were the first spontaneous images. An interesting influence upon some cultures, such as Islam and Judaism, was the religious taboo on the artistic representation of living things. This stifled at birth any tradition of realistic representation of reality. In Islamic art one finds a quite different tradition of geometric design and tessellation, which explores the way in which space can be ordered and divided rather than accurately represented.*

The interesting lesson we learn from these artistic visions is how, until just a few hundred years ago, realism was rather less obvious than it seems to many today. The heavy medieval emphasis upon symbol and schematic representation drove a wedge between raw reality and apprehended reality.†

The move from symbolism to realism brought with it a new attitude towards colour in the post-Renaissance world. Colour plays a central role in the symbolic approach to representation because colours carry meanings. Indeed, they still do; we have only to consider the import of colours in public affairs—in uniforms, religious robes, and national flags. The most heavily symbolic are still gold, black, white, and red. Although colour became less important to the realists than line, composition, and perspective, it offered the greatest scope for novelty. Some, like Georges Seurat, invested great effort in the understanding of colour vision and mixing. Seurat's technique of covering a canvas with a multitude of tiny dots of different colour but similar size, illustrates the principle of colour mixing that our television pictures employ. A television screen displays the colours of three kinds of phosphorescent material that glow when struck by the electron beam that is fired down the picture tube. Each of these materials glows with a different colour when struck, and the eye perceives the overall display of coloured points as an integrated colour picture. Since the intensities of the primary colours, like red and blue, are quite low for these materials, a compromise is made and reddish-orange, hazy blue, and yellowy-green colours are used as the building-blocks. Artists can adopt the same approach by painting many small points of colour, which, when viewed from a distance of a few feet, are mixed by the eye to produce a smoothly varying field of colour (Plate 4). When viewed close up the graininess is evident.

---

* Likewise, the particular character of the Christian historical tradition, with its focus upon the crucifixion of Christ, directed Western artists to perfect the static representation of the naked human form. Religious symbolism and the desire to represent an historical event overcame natural modesty with regard to the representation of the naked human form. If, as Enoch Powell claimed, Christ had been stoned, then the artistic representation of motion and human movement would have evolved more significantly than did the representation of the static human body.

† Of course, there was eventually a reaction against the convergent trend towards ever-increasing realism by the cubists and expressionists, and against the domination of colour by symbolism by the fauves and the impressionists. Many reasons for these reactions have been offered. There are those who see them as nothing more than the ubiquitous 'swing of the pendulum', to be found in so many human affairs: as it gets harder to create new and interesting work by continuing to develop in one direction, so the chances of an iconoclastic U-turn become greater. Alternatively, there are those who look for parallels with developments in other spheres of human activity—musical, scientific, social—to fuel artistic deviations from the status quo. Such a parallelist approach still leaves unanswered the question of why a change occurs in any of these activities. Moreover, influences from other spheres are usually hotly denied by practitioners of any art-form.

Seurat exploited this process most literally, but it was used with even greater subtlety by Monet and many other impressionists. By generating a field of colour additively, a contrast is created with the traditional 'subtractive' creation of colour, which mixes pigments of different primary colours. This is called a 'subtractive' method because the pigment does not produce light of the required colour. A blue pigment is so called because it absorbs all the colours in the spectrum of white light other than blue. This means that the laws for the addition of coloured lights are quite different from those for the addition of coloured paints. Red paint will absorb all colours from white light except red; green paint all but green. Hence, if red and green paints are mixed, everything is absorbed and a dull black mixture results. By contrast, the proportion of light reflected by the different colours is rather similar and so it is very hard to produce a bright mixture with a narrow colour range; most mixtures just produce a muddy brown.

▓▓ The mind-benders: distortions of thought and space

Literature expresses itself by abstraction, whereas painting, by means of drawings and colour, gives concrete shape to sensations and perceptions.

Paul Cézanne

The Western picture of the mind, separate from the body, perceiving the outside world alone and undisturbed, underwent its most profound scrutiny at the hands of the eighteenth-century German philosopher Immanuel Kant. In early life, Kant was an enthusiast for a scientific description of the world based upon Newton's laws of motion and gravitation. He made important contributions to the subject of astronomy—proposing a theory for the origin of the solar system—and was content with the common view that there was a real world 'out there' that could be described by our minds. But, despite his early success, Kant developed an increasingly critical attitude towards the nature of human knowledge, and how it is acquired. He recognized that the human mind does something when it processes sense perceptions of the outside world. It organizes information. Our minds could be said to have pigeon-holes, or categories, into which our perceptions of the world have to be squeezed. And so there must exist an irreducible gap between how the world truly is, and our apprehension of it. We can never know the unexpurgated, untranslated 'things in themselves', only an edited—and possibly distorted—version that has been filtered through our conceptual apparatus. Our conception of its nature will be biased by the range of mental images that we can accommodate, as Figure 2.3 parodies.

Kant seized upon this point to undermine all sorts of woolly claims that his contemporaries had been confidently making about the nature of reality, and then used it as the starting-point for his own complex theory of knowledge. Kant

2.3 Pablo Picasso's *Artist and Model, c.* 1932, Cahiers d'Arts.

sees us as observers of the world, who are denied access to the true observer-independent reality—a fact that places each of us firmly at the centre of our own 'little' universe.

Let us consider an example where our minds are torn between two possibilities by the problem of perspective. It was discovered by a Swiss crystallographer, Louis Albert Necker, in 1832. If we stare at the cubes in Figure 2.4, then the sense of perspective that we rely upon to create a good three-dimensional interpretation of the purely two-dimensional image cast upon the back of the retina is confused: there is no unique three-dimensional image that produces this two-dimensional projection.* The brain has constructed two models of a solid cube, each with a different orientation in space, and it flits between the two, offering us both possible perspectives. It is as if there is an advantage in occasional flips to another view of things, just in case the one already chosen is mistaken. Whole artistic movements have grown up exploiting this image-processing ambiguity. Victor Vasarely, and others in the op art movement, have created intricate images that exploit uncertainties in the brain's identification of lines, and its associations between shapes and points, so that

* An acoustic form of this perceptual ambiguity exists with musical chord sequences displaying the *diabolus in musica* phenomenon.

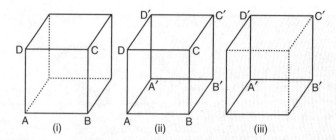

2.4 The Necker cube, with all lines solid, is shown in the centre as (ii). On either side, (i) and (iii) offer alternative visual interpretations of it. On viewing the Necker cube we may see the interpretation (iii), followed soon afterwards by the interpretation (i), followed by rapid shifting between the two as we try to decide whether A or A′ is nearer to us. As Necker first emphasized, the most impressive distinction between (i) and (iii) appears to be the orientation of the cube.

there is constantly changing perspective. The image never appears static. An example of this dynamic art-form is shown in Figure 2.5. It illustrates the impact of our mental categories upon perception: the lines on the page *don't* move whatever your eyes tell you.

Despite the force of Kant's sceptical attitude to the possibility of mind-independent knowledge of the world, there are puzzles. Why are so many people in agreement about so many of the things they see? It seems that humans share many identical categories of thought. Why does our mental picture of the world remain relatively constant from moment to moment? Is there any reason why our mental categories could not change overnight?

There are two poles of opinion about the relationship between true reality and perceived reality. At one extreme, we find 'realists', who regard the filtering of information about the world by mental categories to be a harmless complication that has no significant effect upon the character of the true reality 'out there'. Even when it makes a big difference, we can often understand enough about the cognitive processes involved to recognize when they are being biased, and make some appropriate correction. At the other extreme, we find 'anti-realists', who would deny us any knowledge of that elusive true reality at all. In between these two extremes, you will find a spectrum of compromise positions extensive enough to fill any philosopher's library: each apportions a different weight to the distortion of true reality by our senses.

We can see that Kant's perspective is worrying for the scientific view of the world. At the end of the eighteenth century there was great confidence in the successes of science in uncovering the secrets of Nature. The triumph of Newton's 'laws' of Nature led to ever more confident assertions that the perfect harmony of Nature's laws, and their accord with human well-being, pointed to

2.5 Op art in the form of Bridget Riley's *Fall*, 1963. Tate Gallery, London.

the existence of a benign legislating Deity. Kant's arguments undermined the force of any argument for the existence of God that appealed to the observed laws as evidence for the anthropocentric design of Nature. Those laws might be imposed upon the world by our mental categories of thought: they do not necessarily reflect the true nature of things. This is not an argument against the existence of God, or even one against the anthropocentric design of Nature. That was not Kant's target—in fact, he was rather sympathetic to the aims of those Design Arguments. Rather, he sought to convince his readers that we cannot use the evidence of our senses, or our thoughts, to draw absolutely reliable conclusions about the ultimate nature and purpose of any 'true reality'.

If Kant had lived in the computer age, he would have said that the mental categories that order basic aspects of our experience of the world, like our intuitions about space and time, are 'hard-wired' into our brains. Picking out these hard-wired features of the brain is not easy. Kant viewed our conception of space as being one of these innate, unalterable mental categories. It was not something that we learned by experience. It was a ground of our experience. In choosing our perception of space in this way, Kant was influenced by the abiding belief in the absolute character of Euclidean space. This is the geometry of lines on flat surfaces which we learn at school. It is characterized by the fact that if we form a triangle by joining three points by lines of shortest length then the sum of the three interior angles of the triangle is always equal to 180 degrees (Figure 2.6).

The discovery of such truths, and others (like Pythagoras' theorem for right-angled triangles) led philosophers and theologians to believe in the existence of absolute truth, and in our ability to discern (at least) part of it. The formulation and presentation of medieval theology is not dissimilar to the style of Euclid's classical *Elements* of geometry. This is no accident. It witnesses to a desire to see theological deductions accorded the status of theorems of mathematics. Euclidean goemetry was held up as a piece of absolute truth about the nature of the world. It was not merely a piece of mathematical reasoning about a possible world; it showed how reality truly was. It underpinned the belief of theologians and philosophers that there was reason to believe in the existence of absolute truth. Moreover, we had discovered it, and understood it. Thus we could have confidence in our ability to appreciate, at least partially, absolute truths about the Universe. It is against this background that Kant's choice of Euclidean geometry as a necessary truth about reality must be seen. Unfortunately, it turned out to be a bad one. Not long afterwards, in the mid-nineteenth century, Karl Friedrich Gauss, Johann Bolyai, and Nikolai Lobachevskii all discovered that there can exist other, logically consistent geometries that differ from Euclid's conception. These 'non-Euclidean' geometries describe the properties of lines and curves on a surface that is not flat, and where triangles constructed from the shortest lines between three points do not have interior angles that add up to 180 degrees (see Figure 2.6).

Kant believed that our apprehension of Euclidean geometry was inescapable because it was pre-programmed into the brain. We know this is not true. Not only can we readily conceive of non-Euclidean geometries but, as Einstein first proposed and observations have since confirmed, the underlying geometry of the Universe is non-Euclidean. But it is only over astronomical distances that this deviation from Euclid's rule shows up. It is a property of all curved surfaces that they look flat when viewed locally over sufficiently small regions. The Earth's surface is curved, but seems flat when we sail short distances. Only when we observe accurately over large distances does the curvature of the horizon become evident as in Manet's famous seascape *Boats* that was painted in 1873 (Figure 2.7).

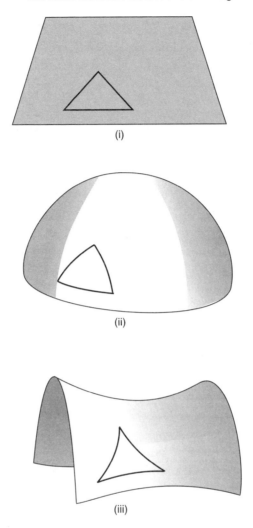

2.6 (i) A Euclidean triangle on a flat surface and two non-Euclidean triangles on (ii) closed and (iii) open curved surfaces. On the flat surface the interior angles of the triangle sum to 180 degrees. On the closed surface they sum to more than 180 degrees; on the open surface to less than 180 degrees. On each surface the definition of a 'straight line' joining two points is the shortest distance between them that lies on the surface.

This geometrical discovery dealt something of a blow to the confidence of theologians and philosophers in the concept of absolute truth. It gave credence to many forms of relativism that are now familiar to us. Books and articles appeared exploring the consequences of the non-absoluteness of any particular system of assumptions for codes of ethics, economic systems, and attitudes

2.7 Edouard Manet's *Boats at Berck-sur-Mer* of 1873; in the Cleveland Museum of Art.

towards non-Western cultures. Whereas there had previously been reason to believe, by analogy with the incontrovertible nature of Euclid's geometry, that there was a 'best' system of values, or for running an economy, and all others were inferior, there was now reason to think again. Later, mathematicians would undermine the grounds of absolute truth even further by showing that not even the rules of logical reasoning that Aristotle had bequeathed us are absolute. As with geometries, so with logics: there are an infinite number of consistent schemes of logical reasoning that can be constructed. There is no such thing as absolute truth in logic and mathematics. The best that one can do is talk of the truth of statements given a set of rules of reasoning. It is quite possible to have statements that are true in one logical system, but false in another.

Much has been written about the impact of the development of non-Euclidean geometry upon artistic images of the world at the beginning of this century. Some have argued that the introduction of new geometries, and the revised conceptions of space and time that emerged through Einstein's theories of relativity, inspired the development of new geometrical art-forms like cubism—although Picasso claimed that no direct artistic inspiration was drawn from the theory of relativity at all, saying that

Mathematics, trigonometry, chemistry, psychoanalysis, music and whatnot have been related to cubism to give it an easier interpretation. All this has been pure literature, not to say nonsense, which brought bad results, blinding people with theories. Cubism has kept itself within the limits and limitations of painting, never pretending to go beyond it.

Those seeking such motivations for new, unconventional, or abstract forms of art may be looking in the wrong place. It cannot be the curved lines and triangles of non-Euclidean geometry that evince such novelty that they inspire Manet to paint a realistic curved horizon in a seascape, or Cézanne and Picasso to distort and depart from the traditional styles of representation.

Non-Euclidean geometries have always been all around us, and were well appreciated by artists long before they were ever recognized by mathematicians.*
One has only to look at a fifteenth-century work like the *Arnolfini Wedding* portrait (Plate 5) by Jan van Eyck to see this. There, the Tuscan merchant Giovanni Arnolfini and his wife, together with their faithful dog, are shown in their home; the entire scene is perfectly reflected in a convex mirror hanging on the wall behind them (shown inset) and the perspective is complicated by the use of more than one vanishing-point. The fact that the logically impeccable system of Euclidean geometry gave plane shapes that could be viewed in a distorting mirror should have suggested that the distorted view was equally consistent as an axiomatically defined geometry, and could have been uniquely created on a plane surface by applying some different set of 'distorted' rules. It is intriguing that the technique of anamorphosis used by artists from the sixteenth century onwards (see the example in Figure 2.8) was also based on such distortions, but the emphasis was entirely upon the fact that the plane 'Euclidean' image can be restored by viewing at an angle or in a suitably curved mirror, rather than upon the logical consistency of the 'non-Euclidean' image.

A revolutionary change in outlook on the world, and its representation, may have been fostered by the general climate of relativism that was encouraged by the discovery that even geometrical truth was not absolute. If there was no reason to believe in absolute mathematical truths about the world, why should there be only one way of painting it, or only one logic to govern our thoughts about it? This general climate of exploring new possibilities, where once there was certainty, was likely to have been more influential than any formalization of geometry that was already, if only unconsciously, well appreciated by artists.

---

* The most impressive examples are the non-Euclidean geometrical constructions in some *Sriyantra*, designs used to assist meditation in various parts of the Indian tantric tradition. Although most *Sriyantra* seals are planar and Euclidean, some examples exist in which the intricate design lies on a regular curved surface. For further details see my *Pi in the Sky* (Clarendon Press, Oxford), p. 76.

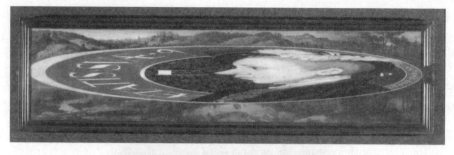

2.8 A sixteenth-century anamorphic portrait of Edward VI by William Scrots, seen (top) face on and (below) obliquely, whereupon the undistorted image is restored.

▓ The inheritors: adaptation and evolution

> Icarus soared upwards to the Sun till the wax melted which bound his
> wings and his flight ended in fiasco . . . The classical authorities tell us, of
> course, that he was only 'doing a stunt'; but I prefer to think of him as the
> man who brought to light a serious constructional defect in the flying
> machines of his day.
>
> Arthur S. Eddington

By the eighteenth century, suspicions were developing that the spectrum of
living things was not fixed. Some transformation of their bodily features and
habits from generation to generation was evidently possible. This could be seen
by looking at the results of selective breeding. It was also becoming clear that
many living species had become extinct. Exotic beasts—mammoths and sabre-
toothed tigers—had left dramatic fossilized remains; by the turn of the nineteenth
century, their study would become a fully fledged science. These facts hinted at
the unsatisfactory nature of a belief that living things were created in perfect
harmony with their environments and with their fellows. There nevertheless
remained the impressive fact that living things appeared to be tailor-made for
their environments. This convinced natural theologians that a form of divine
guidance was operating in the living world, designing creatures to complement
their habitats in an optimal way. Conversely, others argued that the existence of
such a close match between the environmental requirements of living things and
the status quo showed that a grand design existed—and hence there must be a
Grand Designer. Other forms of the Design Argument existed, but were rather
different. They appealed, not to the remarkable interrelationships between
aspects of the environment and the functioning of living organisms, but to the
wonderful simplicity and universality of the laws of Nature that governed the
motions of the Earth and the planetary bodies. Such arguments tended to attract
religiously minded physicists and astronomers rather than biologists.

The first attempt to develop a theory that explained the striking compatibility
between organisms and their environments by appealing to changes that caused
the two to converge over time was made by the French zoologist Jean Baptiste de
Lamarck (1744–1829). Like the natural theologians, Lamarck started from the
assumption that organisms are always well adapted to their surroundings. But
unlike them, he recognized that, because environments change, so must the
organisms if they are to remain in a state of adaptation. Lamarck believed that
environmental changes would lead organisms to learn new behaviours, or
develop anatomical changes, which would be reinforced by repeated exercise. By
contrast, their counterparts that fell into disuse would gradually wither
away. Any structural or behavioural changes induced by the new environmental
conditions would maintain a state of adaptation that could be passed on by

inheritance. Underlying the whole process was a belief that living things tend to evolve towards the most harmonious and perfect structural forms.

In Lamarck's scenario, any changes in the environment determine the evolution of living things directly. As trees grew taller, so giraffes would need to develop longer legs, or necks, in order to continue to feed on their leaves. If a mineworker developed larger muscles by lifting heavy loads, then his muscular frame would be inherited by his children. This, of course, is the type of reasonable supposition that finds its place in folklore; it was already an old notion in Lamarck's time. It was, therefore, not implausible.

Darwin's theory of evolution by natural selection differed from Lamarck's. It abandoned the unwarranted assumption that organisms took their marching orders from the environment as if tied to its changes by some invisible umbilical cord, or directed by an unseen hand. Darwin realized that the environment was an extremely complicated construct, consisting of all manner of different influences. There is no reason why its vagaries should be linked to those that take place within an organism at all. He saw that when changes occurred within an environment, all that resulted was that some organisms found themselves able to cope with the new environment, while others did not. The former survived, passing on the traits that enabled them to survive, and the latter died out. In this way, those features that aided survival, and could be inherited, were preferentially passed on. The process was called 'natural selection'. It cannot guarantee that the next generation will survive; if further changes occur in the environment that are so dramatic that no resident of it can cope with them, then extinction may follow. The essence of this picture is that the environment simply presents challenging problems for organisms, and the only resources available for their solution are to be found in the variation of capabilities among a breeding population. If the environment changes over a long period, then the preferential survival of those members of a species having the greater measure of attributes that best fit them to cope with the environmental changes will result in a gradual change in the species. As a result, new species may, in effect, emerge. The survivors will be better adapted, on average, than their unsuccessful competitors; but there is no reason why their adaptations should be the best possible when judged by some mathematical standard of structural or functional efficiency. Perfection could be a very expensive luxury, and quite impossible in an environment that is constantly changing.

The contrast between Lamarckian evolution and Darwinian evolution is clear. Whereas Lamarck imagined that organisms would produce adaptations in response to the environmental problems they encountered, Darwin saw organisms producing all sorts of traits, initially at random, before there was any need for one of them. No unseen hand is at work, generating only those variations that would be needed to meet future requirements. Darwin called this process 'evolution by natural selection'. It was also discovered independently by Alfred

Russel Wallace. When Darwin published the detailed evidence in support of this proposal in 1859, he did not know how the variation of traits could arise in living things, or how specific traits could be passed on to offspring, thereby perpetuating those that could cope with the existing environment. The work of Gregor Mendel, performed between 1856 and 1871, uncovered inheritable factors (which we now call 'genes') that pass on organic information from one generation to the next. Whereas one might have thought that inherited features would always be just an average of an organism's parents, this turned out not to be the case. Specific features could be inherited undiluted, or even stored unexpressed, only to appear in subsequent generations. During the twentieth century, Mendel's pioneering ideas have developed into the subject of genetics, and then gave rise to molecular biology, which is dedicated to elucidating how genetic information is stored, transmitted, and expressed by DNA molecules. The merger of the concept of evolution by natural selection with the insights into the means by which genetic information is stored, expressed, and inherited by living things, has become known as the 'Modern Synthesis'.

We believe that Darwinian evolution has just three requirements:

- The existence of variations among the members of a population. These can be in structure, in function, or in behaviour.
- The likelihood of survival, or of reproduction, depends upon those variations.
- A means of inheriting characteristics must exist, so that there is some correlation between the nature of parents and their offspring. Those variations that contribute to the likelihood of the parents' survival will thus most probably be inherited.

It should be stressed that under these conditions evolution is not an option. If any population has these properties then it must evolve. Moreover, the three requirements could be met in many different ways. The variations could be in genetic make-up, or in the ability to understand abstract concepts; the mechanism of inheritance could be social, cultural, or genetic. In addition, although the initial source of variation may contain an element that is independent of the environment, there will in general be a complicated interrelationship between the sources of variation and the environment. An outside influence upon the environment may lead to the preferential survival of individuals with particular traits, but those members may have a particular influence upon the subsequent development of the environment. Moreover, the concept of an organism's environment is not entirely unambiguous, for it includes other organisms and the consequences of their activities. Only if the environment is extremely stable will this complicated coupling between organisms and environments be of little importance. Later in this chapter, we shall see that there do exist highly constraining environments that are not altered by their inhabitants.

What is the long-term result of evolution by natural selection? On this question, opinion is divided. Some maintain that the evolution of a sufficiently complex system will never end. All species will continue to change even though their relative fitness remains the same. This state of affairs has been coined the 'rat race'. Alternatively, evolution could approach an equilibrium state in which each organism displayed a suite of traits and behaviours (called 'an evolutionarily stable strategy') from which any deviation would lower the probability of its survival. In this second picture, evolution could cease in an unchanging environment, or in one where all the environmental changes were innocuous. Attempts to investigate which of these long-term scenarios should arise in general have found that it is necessary to consider separately those traits that are subject to some overall constraint. These constraints might be structural—you can't carry items of food that are heavier than a certain weight without collapsing; you can't run faster than a certain limit; you can't grow too large and still fly. These constraints place definite limits upon rat races in certain directions. By contrast, unconstrained features can increase or decrease indefinitely without compromising other abilities. Eventually, unconstrained features of organisms tend to end up in rat races with other species, while those constrained by negative feedback end up in an equilibrium that is characterized by an evolutionarily stable strategy.

From this outline of the theory of evolution by natural selection, it might be too hastily concluded that all the traits and behaviours displayed by living things must be beneficial adaptations to some aspect of the natural environment, or that they must optimize the chances of survival in the presence of competitors relying upon the same resources. There is a danger of turning evolutionary biology into a 'just so story' if we merely presume that all aspects of living things must be the optimal solutions of particular problems posed by the environment. Unfortunately, the situation is not so simple. Although well-defined structural problems are often posed by an environment, changes can occur that are not governed by natural selection. Changes can occur in a population because of purely random fluctuations in the genetic make-up of organisms. If small numbers of two species differ only very slightly in their degree of fitness, then it is possible for the species that we would judge to be fittest, on the average, to become extinct as a result of some small chance variation in its genetic make-up outweighing the systematic trend created by natural selection. When a population is small it is especially susceptible to undue influence by the genetic make-up of its original members (the 'Adam and Eve effect'), and this can outweigh the influence of natural selection. To complicate things further, some genetic variations are selectively neutral in the environments in which they arise and so will not be subject to selection. They might simply be side-effects of the organisms becoming larger or smaller, for instance. Similarly, there may be different strategies that offer indistinguishable advantages to the organism. That is, there can exist different, but equally effective, solutions to the same problem; the fact that one was chosen rather than the other might be due to an 'Adam and Eve effect', or just to an accidental initial choice. Perhaps, for example, there is no adaptive advantage to having our hearts on the left-hand side of our chest; the right seems just as good. Finally, a trait might be hard to interpret correctly as an adaptation, because a single genetic change might express itself as two different features of the organism. One might be advantageous to survival; the other disadvantageous. If the net effect is advantageous, then the second, negative trait can still persist in future generations. Organisms are packages of behaviours, some of which are advantageous, some neutral, others disadvantageous. What determines their likelihood of survival is the overall level of fitness that they bestow in relation to that possessed by competitors in the same environment.

Thus, if an organism displays a particular behaviour, or possesses some structural feature, it does not necessarily mean that this is the optimal behaviour, or structure, required to meet some environmental problem. It may be, as for instance is the case with the hydrodynamic profiles of many fish; but in other instances, as for example when considering why camels come with one hump or two, there may be no such optimal adaptation at all. Nature is extremely economical with resources: profligate over-adaptations to meet one challenge will raise the likelihood of inadequate adaptation elsewhere. Also, behaviours can be highly adaptive without arising from selection. For instance, it is highly adaptive to return to the ground after you jump in the air, but this occurs because of the law of gravitation; it has nothing to do with selection.*

If, however, despite all these caveats, one wants to provide explanations for complex coordinated structures, it is to natural selection that one should look first. Random drift, or vagaries of the initial situation, may alter simple behaviours for a period, but they are not going to provide plausible explanations for intricate living systems of great complexity and stability.

Our actions are not predetermined by the results of the natural selection principle. Ironically, our genetic make-up has enabled us to grow big enough, and develop brains complex enough, to display consciousness. Genetic information alone is insufficient to specify the nature and fruits of human consciousness. Yet,

---

* It is a surprisingly common misapprehension to believe that Nature produces perfect adaptations, presumably because it so often produces very good ones. A similar assumption appears in the argument by Roger Penrose, in *The Emperor's New Mind*, that Gödel's theorem prevents the human mind from being a computational algorithm, *if it is infallible*. I believe, however, that the correct conclusion to draw from this argument (ignoring other objections) is that, because the mind is not a perfect logical device, Gödel's theorem does not tell us anything about limitations in its capabilities. There is no reason why natural selection should endow us with brains that are infallible. Our thinking processes display evidence of all sorts of inconsistencies. Linguistic ability—which is far more impressive than mathematical ability, and of far greater adaptive importance—certainly gives no evidence of being a perfect logical system.

from it flows a vast edifice of individual and social structures, whereby most human actions, and many of the most significant parts of the human environment, derive. By using signs and sounds to transmit information we have been able to by-pass the slow process of natural selection, which is constrained by the lifetimes of the individual members of the species. It is also limited to transmitting generalities, rather than specific items of information about the local geography, the weather, the places to find food, and so forth. Of course, the possession of brains sophisticated enough to learn from experience, rather than merely to respond to genetic programming, does not come cheaply. It requires a vast investment of resources compared with just evolving instinctive genetic responses. It also runs the risk of error and misjudgement in a way that the ingrained instinctive reactions will not, unless the environment changes with unexpected suddenness. With imagination comes risk; but the benefits more than compensate. In a precarious, rapidly changing environment the only way to render survival probable is to predict what might occur, and to plan for a variety of alternatives. We have the capacity to change our behaviour, and to respond to debilitating changes in the environment (by not using CFCs in aerosols, for example). These behavioural changes are not genetically inheritable; none the less, we are able to pass on this information, in written or audio form, so as to by-pass the long time-scales required for genetic inheritance. Moreover, these methods of information transfer offer possibilities for correction and continual revision in the light of changing circumstances and widening knowledge. The pen is indeed mightier than the sword.

After Babel: a linguistic digression

> There could be staccato talk without thought.
> There had to be thought before structured thought.
> Once established, structured talk could be mastered with less thought.
> Once mastered, structured talk makes for more thought.
>
> Florian von Schilcher and Neil Tennant

There is one live area of enquiry where the dilemma of instinct versus learned behaviour is central: the origin of language. Language is so fundamental to our conscious experience that we cannot conceive of its absence. Without language we are trapped. Much of our conscious thinking feels like silently talking to ourselves. But what is the origin of language? There are two poles of opinion and much in between. At one extreme is the view that our linguistic and cognitive abilities are all latent within us at birth, after which they gradually unfold on a time-scale, and with a logic, that is genetically and universally pre-programmed. That programming is part of what defines a human being. At the other extreme we find a belief that the infant mind is a blank sheet upon which knowledge will be inscribed solely through interaction with the world. The first of these views about the origin of language has been explored and developed most extensively by the American linguist Noam Chomsky, who first promulgated it, in the face of much opposition from anthropologists and social scientists, in the late 1950s. The counter view, that our mental appreciation of the world is entirely created by our interaction with it, is often associated with the Swiss psychologist Jean Piaget, who attempted to place it upon a firm foundation by performing extensive studies of learning processes in young children. One of Piaget's central interests was in the process by which children come to appreciate mathematical, geometrical, and logical concepts through manipulating toys that carry concrete information about these abstractions. Simple notions like equality, one quantity being larger or smaller than another, the invariance of objects when moved, and so forth, are extracted from the world by playful experience. A model railway, for example, endows an understanding of logic and geometry, because its construction requires the assimilation of the rules governing the fitting together of the pieces of track. Although Piaget's approach rings true with regard to many aspects of our early learning experience, the acquisition of linguistic skills confronts it with a number of striking facts that Chomsky used to support his view that language is an inbuilt instinct.

Although children are exposed to the structure of language—its syntax and grammar—only at a superficial level, they are able to carry out many complicated abstract constructions. The average five-year-old's exposure to language is insufficient to explain his or her linguistic proficiency. Children can use and understand sentences that they have never heard before. No matter how poor they may be at other activities, able-bodied children never fail to learn to speak. This expertise is achieved without specific instruction. The amount of environmental interaction that they experience is insufficient to explain their linguistic proficiency. Children seem to develop linguistic proficiency most rapidly between the ages of two and three irrespective of their exposure levels. Attempts to learn foreign languages by older individuals do not meet with the same success, nor do adults respond to the same educational process. The sponge-like learning ability of a child appears to turn off at an early age.

Language seems to be an ability that is potentially infinite in scope. How can it arise solely from very limited and necessarily finite experience of the world? A detailed study of the structure of human languages has revealed a deep unity in their grammatical structures to an extent that a visitor from outer space might, at some level, conclude that all humans speak different dialects of the same language.

For Chomsky, language is a particular cognitive ability innate to humans. Our brains contain genetically programmed neural 'wiring' which predisposes the learner to perform the steps that lead to language. This initial 'hard-wiring' of the brain is something that members of our species share. When we are first exposed to an environment in which a language is being spoken, it is as if certain parameters in that built-in program are then fixed, and the program then runs upon the raw vocabulary, grammar, and syntax of the language that we hear. The scope, and level of sophistication, that will result from this process will vary from person to person and will be very sensitive to slight variations in experience. This is why infants adapt to and assimilate language so easily. At root, Chomsky argued that language is not a human invention. It is innate to human nature, just as jumping is innate to a kangaroo's nature. But what is innate is a type of program, which develops in response to external stimuli. How that development takes place is a subject for much controversy and research.*

Chomsky views a child's language acquisition as just another one of the many pieces of genetic pre-programming that equips it to pass from childhood to puberty and adulthood. Prior to his proposals, linguists had focused attention upon building up the grammars of as many human languages as possible (almost three thousand are known). Chomsky turned things upside down. Starting with the assumption that the mind is in possession of an unknown 'universal grammar', which has variable parameters that can be set in different ways by different languages, the quest was to uncover the underlying universal grammar from studies of the particular languages that arise from it. Chomsky noticed that we have an intuitive feel for the formal structure of language that is independent of its meaning. He offers us a sentence 'Colorless green ideas sleep furiously'.† We see this as a meaningless piece of English, but we sense that its grammar and form seem right. The categories of thought that delineate form can exist independently of the need to deal with meaning. It is these formal categories that Chomsky saw as the key to language, and his research programme was devoted to isolating the basic formal ingredients that constitute the universal grammar behind all languages.

Piaget presents human intelligence as something that processes information from the outside world, and gradually constructs a model of reality that becomes more sophisticated as we pass through childhood. He appeals to this interactive process as the basis for the acquisition of all our cognitive skills. By contrast, Chomsky seems to deny this active role for the mind, regarding it as a passive receiver of information. The infant does not receive a once-and-for-all impression

---

* Not much has changed since 1866 when the issue of the origin of language was generating so much unfounded speculation that the Linguistic Society of Paris banned its discussion.

† Needless to say this provoked attempts to inject Chomsky's example with contextual meaning. John Hollander's verse 'Coiled Alizarine' is dedicated to Noam Chomsky:

> Curiously deep, the slumber of crimson thoughts:
> While breathless, in stodgy viridian,
> Colorless green ideas sleep furiously.

of how things really are, but fixes the parameters of some pre-existing program in the mind. Our linguistic pre-programming is unique for its purpose, rather than part of some more general programming for problem-solving of all kinds, as Piaget claimed. It is this last claim that made Piaget's position hard to defend. If language acquisition is just another part of our developing problem-solving ability, why is it so distinctive in practice? We have little trouble in learning all sorts of other procedures and acquiring other skills right up until middle age, and beyond; but our instinctive language-acquisition skills do not persist beyond early childhood. After they have learned their native language, by setting the 'switches' in their innate universal program, talented linguists are distinguished by their ability to change the settings and learn other languages—although they do not learn them in the same manner in which an infant acquires its first language.

If we assume that our minds do possess some sort of hard-wiring for language acquisition, it is appropriate to ask if we can narrow down the nature of that hard-wiring any further. The linguist Derek Bickerton has suggested that we are not just hard-wired with a universal grammar and adjustable settings that become fixed by hearing language. Instead, we are actually hard-wired with some of those settings already fixed. They remain like that until overwritten by the language that the child hears spoken in its environment. What is interesting about this view is that it allows some tests to be carried out. If the child grows up in a culture where the spoken language is a primitive mixture of pidgin speech, then the initial settings will not get overwritten and will persist. There is evidence that the initial settings are for a simple creole linguistic form. Typical errors of grammar and word ordering, like double negatives, persist among young children, and are characteristic of the creole form. Speakers thus revert to innate creole grammars if they have not been exposed to a local grammar that resets their linguistic 'switches' to the new form. If the child hears no systematically structured languages, but grows up amidst a collection of unstructured pidgin languages, then the original creole-like settings will tend to persist and become harder to change with the passage of time.

Finally, we might add that Chomsky appears to have an ambiguous attitude towards the origin of our universal grammar. Although there is strong evidence that language is instinctive, and not a learned behaviour, we must still explain the origin of the universal grammar, determine whether it is one of many possibilities, and uncover the step-by-step process by which it evolved from more primitive systems of sounds and signs. As yet, there has been little progress in addressing those problems. In general terms, we can see that language is adaptive: it confers huge advantages upon its practitioners. Once it became a genetic possibility, there would be enormous pressure for its propagation, and selection for its improvement. The precise sequence of evolutionary steps is, however, likely to be quite complicated to reconstruct, because language requires a combination of anatomical designs to coincide with mental programming in order to be effective.

▓▓ A sense of reality: the evolution of mental pictures

> Human beings are what they understand themselves to be; they are composed entirely of beliefs about themselves and about the world they inhabit.
>
> Michael Oakeshott

Kant's view that our conception of the world is separated from its reality by our cognitive apparatus must be modified in the light of what we have learnt about the evolution of organisms and environments. Cognition, too, is subject to evolution. Plato first recognized that 'observation' involves doing something. Our senses are already in place before they receive sensations. But this potentially profound insight was followed up by a less convincing claim that our instinctive knowledge of things arose because we possessed foreknowledge of blueprints for every particular thing we might encounter in the world. This is an extremely inefficient way to design a system. Kant was more economical: he did not want to endow us with hard-wired knowledge of every particular thing, just of general categories and modes of understanding. Using these categories we could construct conceptions of things, as we might construct buildings from bricks. These innate categories of thought were supposed to be universal to all unimpaired humans. But why should this be? Since Kant could not say where these mental pigeon-holes came from, he could not be sure that they would not suddenly start to change, or that they would not differ from one person to another.

There is one vital truth about the nature of things that we now appreciate, but Kant did not. We know that the world did not appear ready-made. It is subject to inevitable forces of change. This view of things began to emerge during the nineteenth century. Astronomers began to describe how the solar system might have come into being from an earlier, more disordered state; geologists began to come to terms with the evidence of the fossil record; physicists became aware of the laws governing the changes that can occur in a physical system with the passage of time. But the most significant contribution was Darwin's, and it has become clear that it has important things to teach us, not merely about fruit flies and animal habitats, but about Kant's deep questions concerning the relationship between reality and perceived reality.

A consideration of the evolutionary process that has accompanied the development of living complexity dispels some of the mysteries of why we share similar categories of thought: why we have many of the categories that we do, and why they remain constant in time. For these categories have evolved, together with the brain, by the process of natural selection. This process selects for those images of the world that most accurately model the character of its true underlying reality in the arena of experience where adaptation occurs. Evolutionary biology thus lends support to a realist perspective about an important part of the world: that part of which the correct apprehension is advantageous. Many of those apprehensions do not merely give us advantages over others who possess them to a lesser extent: they are necessary conditions for the continued existence of any form of living complexity. Minds that came spontaneously into being with images of the world that did not correspond to reality would fail to survive. Those minds would contain mental models of the world that would be rendered false when confronted by experience. Our minds and bodies express information about the nature of the environment in which they have developed, whether we like it or not. Our eyes have evolved as light-receptors by an adaptive process that responds to the nature of light. Their structure tells us things about the true nature of light. There is no room for a view that all our knowledge of light is nothing more than a mental creation. It is precisely because it is a creation of our minds that our knowledge of light contains elements of an underlying reality. The fact that we possess eyes witnesses to the reality of that something we call light.

Although we do not know whether we are alone in the Universe, we are certainly not alone on the Earth. There are other living things with a variety of levels of 'consciousness' reflected by the sophistication of the mental models that they are able to create of the world around them. Some creatures can create a model that can simulate the future under the assumption that it will develop in an identical fashion from similar circumstances in the past. Other creatures, like crocodiles, lack this ability to link past, present, and future, and live in an eternal present. All plants and animals have encoded a model, or embodied a theory, about the Universe that equips them for survival in the environment they have experienced. Those models vary greatly in sophistication. We know that an ant is genetically programmed to carry out certain activities within its colony. It possesses a simple model of a little piece of the world. Chimpanzees possess a far more sophisticated model of reality, but we know that it is none the less a drastic abbreviation of what can be known about the world. We could place a chimpanzee in a situation that would be beyond its ability to comprehend successfully—at the controls of a flight-simulator, for example. While our own mental pictures of the world are more sophisticated than those of any other terrestrial life-form, they are none the less incomplete. Remarkably, they are complete enough to recognize that they must be incomplete. We know that when we look at a chair we receive only some of the information about it that is available to observers. Our senses are limited. We 'see' only some wavelengths of light; we 'smell' only a range of odours; we 'hear' only a range of sounds. If we see nothing, then this does not mean that nothing is there. The extents of our senses, both quantitatively and qualitatively, are also the results of a selection process that must allocate scarce resources. We could have evolved eyes that were thousands of times more sensitive, but that ability would need to have been paid for by using resources that could have been used elsewhere. We have ended up with a package of senses that makes efficient use of the resources available.

Despite the power of the evolutionary underpinning of a broadly realist view of things, we must be careful not to claim too much. We have already seen that some features of organisms can exist as harmless by-products of adaptations for other purposes. The same holds for our images of reality. Moreover, we find ourselves in possession of an entire collection of abilities that have no obvious selective advantage. Wallace, the co-discoverer of the theory of evolution, failed to recognize this subtlety, and concluded that many human abilities were inexplicable on the basis of natural selection. But Darwin was better able to appreciate the fact that we are bundles of abilities, outdated adaptations, and innocuous by-products. The distinguished theoretical biologist John Maynard Smith argued that

It is a striking fact that, although Darwin and Wallace arrived independently at the idea of evolution by natural selection, Wallace never followed Darwin in taking the further step of asserting that the human mind was also a product of evolution . . . [Stephen Jay Gould suggests that this was] . . . because Wallace had a too simplistic view of selection, according to which every feature of every organism is the product of selection, whereas Darwin was more flexible, and recognised that many characteristics are historical accidents or the unselected corollaries of something that has been selected. Now there are features of the human mind which it is hard to explain as the products of natural selection: few people have had more children because they could solve differential equations or play chess blindfold. Wallace, therefore, was driven to the view that the human mind required some different kind of explanation, whereas Darwin found no difficulty in thinking that a mind which evolved because it could cope with the complexity of life in primitive human societies would show unpredictable and unselected properties.

While we can understand how key notions, like those of cause and effect, are necessary for successful evolution by natural selection, it is not so easy to see why mental images of elementary particles or black holes should be underwritten in the same way. What survival value can be ascribed to the understanding of relativity and quantum theory? Primitive humans evolved quite successfully over hundreds of thousands of years without so much as an inkling about these aspects of the Universe's deep structure. But these esoteric concepts are merely collections of much simpler ideas joined together in complicated ways. Those simpler ideas have far wider currency, and are useful in evaluating a vast range of natural phenomena. Our sophisticated scientific knowledge might be seen as a by-product of other adaptations for the recognition of order and pattern in the environment. Artistic appreciation is clearly closely connected with this propensity. But a susceptibility for recognizing patterns and ascribing order to the world is a powerful urge. The abundance of myths, legends, and pseudo-explanations for the world witness to a propensity we have for inventing spurious ordering principles to explain the world. We are afraid of the unexplained. Chaos, disorder, and chance were closely linked to a dark side of the Universe: the antithesis of the benevolent gods. One reason for this is that the recognition of order has passed from having some reward that is beneficial—recognizing food sources, predators, or members of the same species—to becoming an end in itself. There is a satisfaction to be gained from the creation of order, or from the discovery of order. These feelings probably have their origins in an evolutionary past, where the ability to make such identifications was adaptive.

Because our minds and sensibilities have developed in response to a selective process that rewards correspondence to the way the world is, we can expect to find variations in those mental attributes constrained and entrained by some aspects of the underlying structure of the Universe. The environment in which we have evolved goes deeper than the superficial world of other living things. It springs from the laws and constants of Nature that determine the very form and fabric of the Universe. The complexity of our minds and bodies is a reflection of the complexity of the cosmic environment in which we find ourselves. The nature of the Universe has imprinted itself upon us, constraining our sensibilities in striking and unexpected ways.

The care and maintenance of a small planet: cosmic environmentalism

> The theoretician's prayer: 'Dear Lord, forgive me the sin of arrogance, and Lord, by arrogance I mean the following . . .'
>
> Leon Lederman

The evolutionary process ensures that we have become embodiments of many aspects of our environment whose existence is necessary for our survival. But what exactly is this environment? Biologists have long taught us about the manner in which the immediate climate, topography, and available resources determine the conditions under which evolution occurs. In recent years, we have become aware of broader conditions that underwrite any and every form of life on Earth. As human expansion and influence have grown to levels that challenge the stability of the entire terrestrial environment, we have discovered how the origin and persistence of life owes much to an unseen balance that is deep and delicate. Ironically, many aspects of this balance have become known to us only through our unwitting displacement of them. The growth of technological output and its waste products has begun to change the climate of the Earth. By the time we discover whether this is a systematic trend, rather than a short-lived, staccato fluctuation, it might well be too late to do anything about it. Other human activities have spawned waste gases that alter the chemical processes that control the abundance of ozone gas in the atmosphere. As the ozone layer thins, we shall find ourselves prey to an intensity of ultraviolet light that the slow process of evolution did not equip us to tolerate. Depletion of the ozone layer will accelerate damage to human cells, and increase the incidence of lethal skin cancers. Unsuspected influences also come from beyond the bounds of our solar system. In 1992 the world's news media became excited by predictions that, after a near miss this time round, Comet Swift–Tuttle would return on 14 August 2126 and score a direct hit on the Earth: an event that would bring an end to all human life. Indeed, it has been argued that past terrestrial impacts by debris from space have played a major role in the mass extinctions of life on Earth that are inscribed in the fossil record. It is now widely believed that a comet or meteor collision contributed to the mass extinction 65 million years ago, in which the dinosaurs died out. The dust and debris from the impact rose high into the atmosphere, shrouding the surface of the planet from the Sun's rays for a period long enough to kill all the plants on which the food chain rested. Other extinctions occurred at other times. Paradoxically, such catastrophic extinctions may even have been a necessary precursor to our own rapid evolution to sentience, because an enormous increase in the diversity of life seems to blossom as the environment recovers from these catastrophes. By clearing the ecological stage, extinctions take the brakes off the evolutionary process by opening up large numbers of unoccupied environmental niches. A period of rapid diversification ensues before the usual constraints of overcrowding and scarce resources are imposed again.

One can sometimes witness this rapid expansion into vacated niches on a smaller, local scale. A few years ago, the south-east of England was devastated by a surprise hurricane which generated the highest wind speeds ever recorded over the British Isles. In the counties of Sussex and Kent, entire areas of woodland disappeared overnight. Stanmer Woods, on the edge of the University of Sussex, was especially badly hit. One day, I looked out of my window and saw a vast and ancient forest of elm trees; the next, only a bare horizon covered with battered lumber, twisted branches, and fallen leaves. As the timber was gradually burnt or cleared away, the woods looked barren and desolate, but with the passage of time a vast new diversity of flowers, young trees, and bushes has appeared. The disappearance of the trees allowed light to penetrate to ground level, left far more moisture in the ground, and created space for other things to grow. Of course, no species were driven to extinction by the storm, but the way that the woodland has recovered surprisingly quickly from a mass destruction of trees and loss of birds displays a rich diversity that is a microcosm of the whole Earth's recovery from occasional ecological catastrophes, millions of years ago.

To a first approximation, life *is* extinct. More than 99 per cent of all the species that have ever lived have gone the way of the dinosaurs. The constant battle between reproductive success and obliteration has only narrowly favoured the former. Before the extinction of the dinosaurs, mammals had been rather few and far between, and most of them pretty much shrew-like. Soon afterwards, virtually all the present vast diversity of mammals, from mice to elephants, sprang up in just a dozen million years.

The fossil record shows that, if the rate at which diversity seems to have appeared across the Precambrian–Cambrian boundary had continued unabated to the present, then the oceans would contain more than 10^{27} different marine species, instead of the one million or so estimated to exist today. Clearly evolution could go far faster than it does. Presumably it is attenuated by the limited space and resources that exist to support different creatures.

Major environmental catastrophes may be necessary for evolution to reach high levels of diversity and sophistication by a series of relatively rapid steps. If life originates on other crowded worlds, then the emergence of complex life-forms may require a succession of catastrophic events to accelerate the pace of evolution. Without them, evolution may wind slowly down. Safe, uneventful worlds are not necessarily advantageous to the life process: to live in complex ways, you must live dangerously because coping with danger necessitates the evolution of complexity.

If mass extinctions were caused by local events internal to the environment—some sort of disease, for example—then one might imagine that the evolutionary process would produce more offspring with increased resistance to such threats, and extinctions would become rarer and less catastrophic. As a result, the potential for rapid evolutionary change and innovation would be suppressed. Only catastrophic events for which there would be no scope for the systematic genetic evolution of resistance would be able to reset the clock of evolution by huge and unpredictable interventions. The only way in which this cycle can be broken, and large-scale disasters overcome, is by the production of a trait like consciousness that enables information to be transmitted far more quickly than by genetic means.

Seen in this light, it may be that the overall rate of evolution of life on Earth has been significantly—and positively—influenced by events like climatic change or the effects of outside perturbations from space. Of course, if we were to be the next candidates for, say, a mass extinction produced by a cometary impact, we might find it hard to take the long-range viewpoint that uses the word 'positively' to describe these influences. These cosmic encounters are not so improbable that we can completely ignore them. In 1992, we learnt of the threat from Comet Swift-Tuttle. In July 1994, astronomers were given a chance to witness the consequences of a cometary impact on our astronomical doorstep when the fragments of Comet Shoemaker-Levy 9 hit the far side of the planet Jupiter. The energy released by the exploding fragments was millions of times greater than that from the largest terrestrial nuclear explosions. Other recent close approaches to the Earth are known, and have led to serious debate as to how we might best develop defences against such celestial bombardment. Some have proposed that a modified Star Wars technology should be developed with a view to shooting down, or diverting, incoming comets and asteroids when they are in the outer solar system. Others believe that the pursuit of such powerful weapon systems creates greater dangers for humanity than the objects they are designed to shoot down. After all, any technology advanced enough to deflect a small celestial body past the Earth might, in the wrong hands, be capable of diverting it *on to* a specific part of the Earth's surface.

In the absence of catastrophes, our own existence is made possible by the presence of our friendly neighbourhood star: the Sun. Its stability and distance from us ensure that the average terrestrial environment is relatively temperate: cool enough for liquid water, yet warm enough to avoid a never-ending ice age. But the Sun is not unchanging; we know that its surface displays complex out-bursts of magnetic activity that create regular cycles of sunspot activity. No complete explanation exists for these cycles, and their possible influences upon the Earth's climate remain a topic of recurrent speculation. The Sun is not the only star that could play a critical role in the stability of our environment. In 1987, the observation of an exploding star, a 'supernova', in the Large Magellanic Cloud (a nearby 'dwarf' galaxy in the same local group of galaxies as our Milky Way galaxy) excited astronomers all over the world. If it had occurred nearby, in our own galaxy, it could have extinguished all terrestrial life. It is possible that nearby supernovae explosions in the distant past produced radiation that changed the Earth's ozone layer and influenced the course of evolution for the simple marine and reef-based life-forms that were the precursors of later, more complex organisms.

As we contemplate these astronomical hazards we begin to appreciate how hazardous a business is long-term survival in the Universe, both for ourselves and for others. Figure 2.9 shows the likely frequencies of occurrence and energetic consequences of impacts of increasing size.

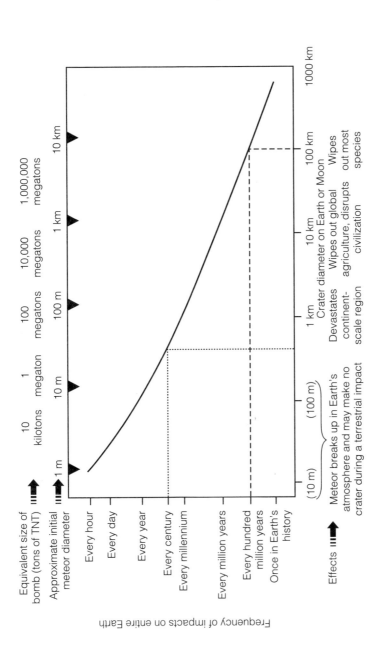

2.9 Constants of Nature. The average frequency of impacts with the Earth's atmosphere by objects of different sizes together with the expected size of the craters created on the Earth's surface and the likely effects.

Based on fig 8.1 in P.D. Ward and D. Brownlee, *Rare Earth*, Copernicus, New York (2000), p.165.

The recognition of these hazards may help us understand deeper mysteries about life in the Universe. Many explanations have been offered as to why we find no evidence for the existence of advanced extraterrestrial life in the nearby Universe. Perhaps we are too uninteresting to be worth a contact; perhaps we are too interesting to disturb; perhaps life requires extremely improbable environments to sustain it. Most likely, I feel, is that life never survives for very long periods. Asteroidal impacts, passing comets, bursts of gamma radiation, all these external hazards are common occurrences. We are shielded from many of them by the planet Jupiter and our large Moon. Without those gravitational shields we would have suffered a string of catastrophic impacts that would have served to continually reset the evolutionary clock. Coupled with the threat to life offered by internal hazards like war, disease, or environmental disaster, we begin to see that it is perhaps not entirely surprising that no one is 'out there' in our part of the Universe.

These topical examples illustrate the risks to the Earth's delicately poised environment that are posed by outside cosmic influences. The environmental factors that have shaped the evolution of the Earth's biosphere may be erratic and sudden in their impacts, beyond the scope of most terrestrial life-forms to survive. Yet past celestial catastrophes are as nothing compared with the interventions that lie in the far future. One day, about five billion years from now, the Sun will begin to die. It will have exhausted its supplies of the hydrogen gas that it burns as nuclear fuel. In its last efforts to adjust to this ultimate solar energy crisis, it will expand and vaporize the inner planets of the solar system before contracting towards a final resting state, only a little bigger than the present size of the Earth. At first, this state will be very hot, but over billions of years the Sun will steadily cool off, leaving a dark cinder, fading steadily into invisibility (see Plate 6). Will humanity by then have found a means of moving elsewhere? It seems unlikely. It is bad enough packing to go on a short holiday with a few other family members. Imagine packing for ten billion, none of whom plan to return!*

For good or ill, the cosmic environment stretches farther and wider than Darwin ever imagined. The structure of the Universe beyond the Earth constrains the environment within which the more familiar processes of biological evolution, adaptation, and cultural development can occur. They place limits upon the diversity that is possible on Earth, and fashion our impressions of the world. By developing an appreciation for the subtleties of our cosmic environment, we can begin to distinguish those features that have emerged by chance

---

* The mathematician Greg Chaitin once told me of a science fiction story in which an American family, who keep themselves to themselves, go away for a camping trip, only to return and discover that the rest of the human race has left the planet—without telling them.

from those which are inevitable consequences of the Universe's deep, unalterable structure.

Gravity's rainbow: the fabric of the world

Damn the solar system. Bad light; planets too distant; pestered with comets; feeble contrivance; could make a better myself.

Lord Jeffrey

Let us step back from the minutiae of biological evolution on Earth, where vast complexity is promoted by a process that we have come to call 'competition', whereby each species actually seeks a niche that will *minimize* its need to compete with rivals. All this co-adaptive interaction between organisms and habitats requires a backdrop. Before genes can be 'selfish', before biological complexity can begin to develop, there must exist atoms and molecules with properties that permit the development of complexity and self-replication; there must exist stable environments; and there must exist sites that are temperate enough for those structures to exist. All these things must persist for enormous periods of time.

Deep within the inner spaces of matter, unseen and unnoticed, exist the features that enable these conditions to be met. Ultimately, it is these things that allow life and all its consequences to flourish on our lonely outpost in the suburbs of a nondescript galaxy called the Milky Way. They do not guarantee life but, without them, all structures complex enough to evolve spontaneously by natural selection would be impossible.

There are four aspects of the Universe's deep structure that combine to underwrite the cosmic environment within which the logic of natural selection has allowed the hand of time to fashion living complexity.

Laws of Nature dictate how the world changes with the passage of time and from place to place. At present, we believe that these laws govern the workings of just four natural forces: gravity, electromagnetism, the weak (radioactive) force, and the strong (nuclear) force. Superficially, these forces appear to be distinct in their ranges, their strengths, and in the identities of the particles of matter that are subject to their inflexible jurisdictions. But as their effects are probed at higher and higher temperatures, they change; their differences melt away, together with many of the problems that have beset our past attempts to understand each of these forces as a single autonomous feature of the Universe. Almost all physicists expect that, ultimately, the four natural forces will be found to be different manifestations of one basic 'superforce', which displays its unity only at very high temperatures. Indeed, such a unification has already been experimentally confirmed for two forces (the electromagnetic and weak forces). It is intriguing to learn that the simplicity of the world depends upon the temperature of the environment. At the low temperatures at which life-supporting biochemistry is possible—at which atoms can exist—the world appears to be complicated and diverse. This is inevitable. The symmetries that hide the underlying unity of the forces of Nature from view must be broken if the complex structures needed for living complexity are to arise. The true simplicity of the laws of Nature is evident only in an environment so close to the inferno of the Big Bang that no complex 'observers' are possible. It is no accident that the world does not appear simple; if it were simple, then we would be too simple to know it.

The three hundred years of success that we have enjoyed by using the concept of laws of Nature to make sense of the Universe have had oblique effects. The concept that the ways of the world are governed by externally imposed 'laws', rather than by innate tendencies within individual things, reflected and encouraged religious belief in a single omnipotent Deity who decreed those laws of Nature. The economy of Nature's laws, their comprehensibility, and their universality have, in the past, all been interpreted as persuasive evidences of a Divine artificer behind the workings of the visible Universe.

In addition to the laws of Nature we need some prescription for the state of the Universe when it began or, if it had no beginning, then a specification of how it must have been at some moment in the past. Fortunately, many aspects of the Universe seem to depend very weakly upon how it began. The high temperatures of the early stages of the Big Bang erase memory of many aspects of the initial state. This is one reason why it is so hard to reconstruct the Big Bang, but it also enables us to understand many (although not all) aspects of the Universe's present structure and recent history without knowing what it was like in the beginning. Some cosmologists believe that it would be best if all memory of the initial conditions had been lost, because then every aspect of the present structure of the Universe might be understood without having to know what the initial state of the Universe was like. Others, most notably James Hartle and Stephen Hawking, have in recent years attempted to pick out a special candidate for the initial state.*

Unfortunately, the part of the Universe that is visible to us, despite being all of fifteen billion light-years across, has arisen by the expansion of a tiny part of the entire initial state. Although some grand 'principle' may indeed dictate the average structure of the initial state of the entire (possibly infinite) Universe, that may not help us to determine the structure of the tiny part of the whole that expanded to become the part of the Universe that is visible to us today.

Besides the laws of change and the initial specification of the Universe, we

---

* This was described in Stephen Hawking's book *A Brief History of Time*, and in my own *Theories of Everything*.

need something else to distinguish our Universe from others that we can conceive. The strengths of the forces of Nature and the properties of the elementary objects that the laws govern, which create the fabric of the Universe, are prescribed by a list of numbers, which we call 'constants of Nature'. They catalogue those aspects of the Universe that are absolutely identical everywhere and at all times.* They include the intrinsic strengths of the forces of Nature, the masses of elementary particles of matter, like electrons and quarks, the electric charge carried by a single electron, and the magnitude of the speed of light. At present we can determine these quantities only by measuring them with ever-greater accuracy. But all physicists believe that many, if not all, of these constants should be fixed in value by the intrinsic logic of an ultimate theory of the natural forces. And, indeed, the correct prediction of these constants might be the ultimate test of any such theory.

Our quartet of forces that underwrite the structure of Nature is completed by information about the way the outcomes of the laws of Nature fall out. A deep subtlety of the world is how a Universe governed by a small number of simple laws can give rise to the plethora of complicated states and structures that we see around us, and of which we ourselves are noteworthy examples. The laws of Nature are based upon the existence of a pattern, linking one state of affairs to another; and where there is pattern, there is symmetry. Yet, despite the emphasis that we place upon them, we do not witness laws of Nature. We see only the *outcomes* of those laws. Moreover, the symmetries that the laws enshrine are broken in those outcomes. Suppose that we balance a needle on its point and then release it. The law of gravity, which governs its subsequent motion, is perfectly democratic. It has no preference for any particular direction in the Universe: it is symmetrical in this respect. Yet, when the needle falls, it must fall in some particular direction. The directional symmetry of the underlying law is broken, therefore, in any particular outcome governed by it. By the same token, the fallen needle hides the symmetry of the law that determined it. Such 'symmetry-breaking' governs much of what we see in the Universe, and its origins can be truly random. It allows a Universe governed by a small number of symmetrical laws to manifest an infinite diversity of complex, asymmetrical states. This is how the Universe can be, at once, simple and complicated. To the particle physicist seeking the ultimate laws of Nature, all is governed by simplicity and symmetry; but for those who try to make sense of the chaotic diversity of the asymmetrical outcomes of Nature's symmetrical laws, symmetry and simplicity are rarely its most impressive manifestations. The biologist, the economist, or the sociologist all focus upon the complexities to be found in the higgledy-piggledy outcomes of the laws of Nature. These outcomes are governed neither

* See my book *The Constants of Nature* for a more wide-ranging discussion.

by simplicity nor symmetry. Hundreds of years ago, natural theologians tried to impress their readers with stories of the wondrous symmetry and simplicities of Nature; now we see that, ironically, it is the departure from those simplicities that makes life possible. It is upon the flaws of Nature, not the laws of Nature, that the possibility of our existence hinges.

It is useful to divide our four factors (laws, initial conditions, constants, and symmetry-breakings) into two pairs. The laws and constants of Nature are features that enforce uniformity and simplicity, while initial conditions and symmetry-breakings permit complexity and diversity. These four factors determine the nature of the cosmic environment. Only if their combination falls within a rather narrow range will it be possible for any form of complexity to develop in the Universe. This range delineates the universes within which life is possible. It displays the conditions that are necessary for life to evolve. None of them is sufficient to guarantee that life will evolve, let alone that it will continue to survive. As we uncover the ways in which the cosmic environment meets the conditions needed for life to evolve and persist, we find that they have unusual by-products. They ensure that many of our attitudes towards the Universe and its contents, together with some of our own creations and fascinations, are subtle consequences of the structure of the Universe.

At first, it seems most unlikely that the Universe, as a whole, could have much influence upon things here and now. We are used to local influences being the strongest. But links can be subtle. Who would have thought that the huge size of the Universe had any role to play in our own existence? In fact, for hundreds of years philosophers have been using the vastness of the Universe as an argument against the significance of life on Earth. But things are not quite as they seem. Life is at root a manifestation of a high level of organized complexity at the molecular and atomic level. Any stable form of complexity must be grounded in combinations of chemical elements that are heavier than hydrogen and helium. The chemical form of life that appears to have evolved spontaneously on Earth is based upon the elements carbon, oxygen, nitrogen, and phosphorus, which can perform all manner of molecular gymnastics in combination with hydrogen. But where do elements like carbon come from? They do not emerge from the Big Bang; it cooled off too fast. Rather, they are produced by a slow chain of nuclear reactions in the stars. First, hydrogen is cooked into helium; then, helium into beryllium; and then beryllium burns into carbon and oxygen. When stars explode as supernovae they disperse these biological elements through space. Ultimately, they find their way into planets, plants, and people. The key to this process of stellar alchemy is the time that it takes to effect. Nuclear cooking is slow. Billions of years are needed to produce elements like carbon, which provide the building blocks for complexity and life. Hence, a universe containing living things must be an old universe. But, since the Universe is expanding, an old universe must also be a large one. The age of the Universe is inextricably linked to its size. The Universe must be billions of light-years in size, because billions of years of stellar alchemy are needed to create the building blocks of living complexity. Even if the Universe were as big as the Milky Way galaxy, with its hundred billion star systems, it would be totally inadequate as an environment for the evolution of life—because it would be little more than a month old.

The large size of the Universe may be inevitable if it is to contain life. But the enormous size and sparseness of the Universe in which living beings find themselves has consequences for their view of the world and of themselves. The separateness of the distant heavenly bodies has tempted some to regard them as divine; for others, a growing awareness of the vastness of space has induced feelings of pessimism and ultimate insignificance. Our philosophical and religious attitudes, our speculative fiction and fantasy, have all developed in the light of extraterrestrial life as merely a distant possibility. Extraterrestrials are rare. One of the reasons is the sheer size of the Universe, and the paucity of material within it. If we were to take all the matter on view in the Universe—all the planets, stars, and galaxies—and smooth it into a uniform sea of atoms, we would end up with no more than about one atom in every cubic metre of space. This is a far closer approximation to a perfect vacuum than we could ever create in one of our laboratories. Outer space is, indeed, mostly that—space. Of course, the local density of matter in the solar system is vastly greater than this average value, because it is packed into dense lumps like planets, meteorites, and mountains. If one thinks about gathering this matter into aggregates, we can see how widely separated the planets and stars, and hence any civilizations they might support, need to be. An average density of ten atoms per cubic metre is the same as placing just one human being (of, say, 100 kilograms mass) in every spherical region of space just over a million kilometres in diameter. It is also the same as placing just one Earth-sized planet in every region with a diameter of a million billion kilometres, and one solar system in every region that is ten times bigger still.

Chronicle of a death foretold: of death and immortality

> While there is death, there is hope.
>
> Anonymous politician seeking higher office

The rate at which the Universe expands, and hence its size and its age, is dictated by the overall density of matter within it, because the density of matter determines the strength of gravity, which decelerates the expansion of the Universe. A universe old enough to contain life must be very large, and contain a very low average density of matter. This connection between the size, the age, and the density of the Universe guarantees that civilizations in the Universe are likely to be separated from each other by vast distances. Any very complicated natural phenomenon that relies upon a sequence of improbable processes will be rare in the Universe, and its rarity will mirror the paucity of matter itself. This is frustrating for those who are keen to communicate with extraterrestrials, but for the rest of us it may be a blessing in disguise. It ensures that civilizations will evolve independently of each other until they are technologically highly advanced—or, at least, until they have the capability of sending radio signals through space. It also means that their contact with each other is (almost certainly) restricted to sending electromagnetic signals at the speed of light. They will not be able to visit, attack, invade, or colonize each other, because of the enormous distances that must be traversed. Direct visits would be limited to tiny robot space-probes that could reproduce themselves using raw materials available in space. Moreover, these astronomical distances ensure that even radio signals will take a very long time to pass between civilizations in neighbouring star systems. No conversations will be possible in real time. The answers to the questions posed by one generation will be received at best by future generations. Conversation will be measured, careful, and ponderous. The cultural insulation provided by the vastness of interstellar, and intergalactic, distances protects civilizations from the machinations, or cultural imperialism, of extraterrestrials who are vastly superior. It prevents interplanetary war, and encourages the art of pure speculation. If one could leapfrog the cultural and scientific progression process by consulting an Oracle who supplies knowledge that it would take us thousands of years to discover unaided, then the dangers of manipulating things that one does not fully understand would outweigh the benefits. All motivation for human progress and discovery might be removed. Fundamental discoveries would be forever out of reach. A decadent and impoverished humanity might result.

If we look back through the history of Western culture we can trace a continuous debate about the likelihood of life on other worlds. Our inability to settle the question, one way or the other, fuelled the speculative debate about the theological and metaphysical consequences of extraterrestrial life. For St Augustine (354–430), the assumed uniqueness of the incarnation of Christ meant that extraterrestrial life could not exist, because there would have been a need for incarnations on those worlds as well. Centuries later, an anti-Christian Deist, Thomas Paine (1737–1809), turned this argument upon its head: he found the existence of extraterrestrials to be self-evident, because there was nothing special about us. Since this state of affairs was incompatible with the uniqueness of the incarnation, he concluded that Christianity was in error. More recently, a science fiction trilogy by C. S. Lewis* explored a third possibility seriously: that

---

* *Out of the Silent Planet, Perelandra,* and *That Hideous Strength.*

extraterrestrial beings were perfect and hence neither in need of redemption or a further incarnation. The Earth was a sort of moral pariah in the Universe.

The point of these snapshots is simply to illustrate how the enormous size of the Universe, and the vast distances that necessarily exist between civilizations, has stimulated particular theological questions and metaphysical attitudes. Although the theological ramifications of extraterrestrial life are largely ignored by theologians who think seriously about modern science, there are still shadows of the ancient debate about the theological aspects of other worlds that place the matter in a new light. Many of the enthusiasts who search for signals from the other worlds have argued that signals from more advanced civilizations would be of enormous benefit to humankind. Frank Drake, the leader of a long-term SETI (Search for Extraterrestrial Intelligence) project, has suggested that contact with advanced extraterrestrials would help humanity to deal wisely with the 'dangers of the period through which we are now passing'. Carl Sagan foresaw the attractive possibility of receiving a message that 'may be detailed prescriptions for the avoidance of technological disaster'. Since we are most likely to hear from the longest-lived societies, these are the ones that are most likely to have passed through crises like the proliferation of weapons of destruction, to have avoided lethal environmental pollution from technological expansion, withstood astronomical catastrophes, and overcome debilitating genetic maladies or social malaise. Taken to its logical conclusion, this line of argument leads one to speculate that we are most likely to receive signals from ultra-long-lived civilizations that have discovered the secret of immortality, because they will tend to survive the longest. Drake claims that

We have been making a dreadful mistake by not focussing all searches . . . on the detection of the signals of immortals. For it is the immortals we will most likely discover . . . An immortal civilization's best assurance of safety would be to make other societies immortal like themselves, rather than risk hazardous military adventures. Thus we could expect them to spread actively the secrets of their immortality among the young, technically developing civilizations.

What is so interesting about all these quotations is their presentation of the goals of a search for extraterrestrial intelligence in a manner that makes them sound like a traditional religion. They seek a transcendental form of knowledge from beings who know the answers to all our problems, who have faced them vicariously, and have overcome them. In so doing they have achieved immortality. Their aim now is to give us that secret of everlasting life.

One might argue that immortality is not a likely end-point for the advanced evolution of living beings. It sometimes appears that the universal legacy of evolution by natural selection is to embody behaviours that, while advantageous to survival in the pre-technological area, inevitably prove fatal later on, when the means for total destruction become available. Or, less pessimistically, perhaps the inevitable spread of life always exhausts the resources available to sustain it. These are two reasons why 'immortals' (even if their existence were compatible with the finite age of a Big Bang* universe), or even civilizations that are millions, rather than just thousands, of years old may not exist in practice—even if they can exist in principle.

Death and periodic extinctions play a vital role in promoting the diversity of life. We have already discussed how the sudden extinction of species allows the evolutionary process to accelerate. In this respect, immortals would evolve more slowly than mortals. Immortality also does strange things to urgency. One recalls Alan Lightman's memorable story† about a world in which everyone lives forever. Its society splits into two quite different groups. There are procrastinators who lack all urgency; faced with an eternity ahead of them, there was world enough and time for everything—their motto, one suspects, was a word like *mañana*, but lacking its sense of urgency. By contrast, there were others who reacted to the unlimited time by becoming manically active because they saw the potential to do everything. But they did not bargain for the dead hand that held back all progress, stopped the completion of any large project, and paralysed society. It was the voice of experience. When every craftsman's father, and his father, and all his ancestors before him, are still alive, then experience ceases to be solely of benefit. There is no end to the hierarchy of consultation, to the wealth of experience, and to the diversity of alternatives. The land of the immortals might well be strewn with unfinished projects, riven by drones and workers with diametrically opposed philosophies of life. With time to spare, time might not have spared them.

Death may be a useful thing to have within the evolutionary process, at least until such time that its positive benefits to the species as a whole can be guaranteed by other interventionary means. Of course, the fact that human death occurs on a time-scale that is short has an important impact upon human metaphysical thinking and, as a consequence, dominates the aims and content of most religions. As we have become more sophisticated in our ability to cure and prevent disease, the death-rate has fallen, and the average human life-span has grown significantly in the richer countries of the world. With this increase in life-expectancy has come a greater fear of death, and a reduced experience of it among close friends and family members. There is much speculation about the

---

* One of the problems with the now-defunct steady-state cosmology of Bondi, Gold, and Hoyle, in which the Universe has no beginning and no end, and maintains the same average properties of expansion, density, and temperature always, was that the Universe should be teeming with life. This argument was put forward by the author and F. J. Tipler in Chapter 9 of *The Anthropic Cosmological Principle* (Clarendon Press, Oxford).

† See *Einstein's Dreams* (Bloomsbury, London).

possible discovery of some magic drug or therapy that will isolate some single gene that results in human death by natural causes. By modifying it, some hope that we would be in a position to prolong the average human life-span. It is, however, very unlikely that the evolutionary process would have given rise to organisms that have a single weak link that dominates the determination of the average life-span. It is much more likely that the optimal allocation of resources results in many of our natural functions wearing out at about the same time so that, on average, there is no single genetic factor that results in death. Rather, many different malfunctions occur at about the same time of life. Why allocate resources to develop organs that would work perfectly for five hundred years if other vital organs never last even a hundred years? Such a budgeting of resources would fail in competition with a strategy that spread resources more evenly among the various critical organs, so that they had similar life-expectancies. There is a story about the late Henry Ford that illustrates the application of this strategy in the car industry. Ford sent a team of agents to tour the scrap-yards of America in search of discarded Model T Fords. He told them to find out which components never failed. When they returned they reported failures of just about everything, except the kingpins. They always had years of service left in them when some other part failed irretrievably. His agents waited to hear how the boss would improve the quality of all those components that failed. Soon afterwards, Henry Ford announced that in future the kingpins on the Model T would be engineered to a lower specification.

It might seem reasonable that our bodies should evolve the ability to repair all injuries and impairments to essential organs, just as they heal mundane cuts and bruises. But this could not be an economical use of resources when compared with the investment that would be required to generate new offspring. As animals age and pass the stage when they can reproduce, genetic resources are not invested to repair them. A strategy with benefits for a young organism, but penalties for an old one, will be superior to one with the same average benefit distributed equally, irrespective of the age of the beneficiary. Moreover, any genes that favour young organisms over old ones will tend to accumulate in the population over long time-scales. Thus, a general decay of our bodily functions and in our capacity for self-repair and regeneration is not surprising.

Of course, were any extraterrestrial signal ever to be received, it would have vast philosophical, as well as scientific, significance. Curiously, the former might well outweigh the latter. For instance, suppose that we received a description of some simple piece of physics or chemistry. This might tell us nothing about those subjects that we did not already know; but, if it were to use mathematical structures similar to our own; if it were to display similar ideas about the structure of the physical Universe—analogous concepts, like constants or laws of Nature—then its impact upon our philosophers would be immense. We would have direct evidence for the existence of a single, lawful structure at the heart of Nature that existed independently of the nature and evolutionary history of its observers. In the realm of mathematics, similar deep revelations might emerge. If our messages revealed the use of mathematics in a familiar form, with emphasis upon proof and the manipulation of infinite quantities, rather than upon experimental mathematics with computers searching for habitual relationships, then we would need to reassess our attitudes towards the idea that mathematics exists and is discovered, rather than merely invented or generated by human minds. We would expect extraterrestrials to have logic; but would it be our logic? Would they have artistic activities like music or painting? Since these are activities that exploit the limited ranges of our human senses, we would not expect to find them in the same form but, as we shall see in later chapters, we might well expect to find particular artistic tendencies. Artistic activities that spring from non-adaptive developments could have almost any form. Those that are modifications or by-products of adaptive behaviours might be a little more predictable. The simple fact of having evidence of a capability to communicate information in specific forms would be quite revealing. Artistic appreciation could even turn into a fascinating predictive (scientific?) activity, attempting, on the basis of some primary evidence, largely technical or scientific, to predict the nature of the artistic activities that might have sprung from them. With regard to language, we might find that the genetic programming that seems to lie at the heart of human linguistic ability is just one way of achieving a loquacious end; or, we might find our extraterrestrial interlocutors displaying grammatical programming of a sort tantalizingly similar to our own. Discoveries of this sort would be much more significant than some piece of undiscovered physics or metallurgy that terrestrial physicists might be able to discover for themselves in the future. The things that we would learn about the uniqueness of our concepts, languages, and other modes of description would be things that we could *never* learn without access to an independent extraterrestrial civilization, no matter how far we advanced our own studies.

Let us return to our discovery that the Universe is not only big, but has to be big in order to contain things complicated enough to be called 'observers'. As the centuries have passed astronomers have steadily increased their estimates of the size of the Universe. The responses to this enlarged perspective have been twofold. There have been those who have sought comfort in their belief that, despite our physical insignificance in space, our position was none the less privileged. We were the object of creation; if not in a central position, then certainly of central cosmic interest. By contrast, there were those who despaired at our position in a scheme of things that seemed to care not one wit for our past, our present, or our future. In the early years of this century there were those who saw the impending heat death of the Universe as a final curtain, bringing to an inglorious end all that we value and seek to pass on. Their frustrations still echoes in the words of those like Steven Weinberg, whose popular account of the expanding Universe led him to exclaim that 'the more comprehensible the Universe becomes the more pointless it seems'. Whole movements in theology and 'process' theology grew up in response to the picture of a Universe running down like a great Victorian engine, succumbing to the doctrine of the second law of thermodynamics, which preaches only the inevitability of bad turning to worse. Process theologians developed the concept of an evolving God who does not know all that the future holds. Even today, one finds a sharp distinction between theologians who regard the presence of time, and the flow of events, as being of vital theological importance, and those who, like many modern cosmologists, see the future as already laid out and determined because the whole of space and time must just *be* there.

The point of discussing these two opposing responses to the size of the Universe, and to our incidental position within it, is not to persuade the reader of the correctness of either one of them. Rather, it is to show that these philosophical and theological notions are consequences of the nature of the Universe in which we find ourselves. If the Universe were significantly different; if, somehow, it could be very small, and teeming with other life-forms, who were readily contactable, then our list of important philosophical and theological questions would be very different, and our image of ourselves would have little in common with our present views. We feel like the Universe's only child, and that feeling has many consequences.

These considerations alert us to the snare of believing that all that matters is rational scientific development, and of judging the advancement of hypothetical extraterrestrials solely in terms of their technical progress. The consequences of evolutionary adaptation to unusual environments can be entirely unexpected, and the emergence of consciousness seems to produce unpredictable dual uses of skills that were evolved to meet challenges that no longer exist. Moreover, adaptations that are very successful in the short term can turn out to have lethal long-term consequences—as we have discovered with regard to industrial pollution of the Earth's atmosphere and environment. One way of looking at human thinking is to see it as a progression towards rationality: everything else is like a computer virus in the brain. But this is very hard to justify. Rationality is not much in evidence in the history of conscious life on Earth. On the other hand, mystical, symbolic, and 'religious' thinking—all those ways of thinking that the rationalist would condemn as 'irrational'—seem to characterize human thinking everywhere and at every time. It is as if there were some adaptive advantage to such modes of thinking that offers benefits that rationality cannot provide. How could this be? Even if we could establish beyond all doubt that one set of religious views was correct, this would not explain the phenomenon, because human religious belief has been directed at countless deities, accompanied by a multitude of different rituals and allied beliefs. The existence of one true religion does not help to explain the profusion of other religious beliefs. One possibility is that rationality breeds caution; irrationality, emotional fervour, and blind belief do not. In a world where hostile conflicts were common and a matter of life or death, too much rationality might not be helpful. The fearless zealot who feels guided by supernatural powers is a difficult opponent to overcome. If you believe that your territory is the abode of gods you will defend it more passionately than if it is merely your home. Rationality is undoubtedly advantageous when you have lots of information to apply it to. But when your understanding of things is fragmentary, and requires considerable interpolation to build up a wide-ranging view, it may not be as effective as uninhibited boldness. Would you have embarked on voyages of discovery knowing what we now know about geography and weather conditions in the Atlantic Ocean? The questing spirit of the explorer and the self-sacrifice of the heroic soldier offer clues to the nature of this side of the human psyche. Logically, it should not exist; but perhaps the advantages that irrational, speculative, and religious beliefs offer through their ability to spur us to actions with positive consequences are significant enough to account for our propensity towards their adoption. Extraterrestrial robots who were completely rational might evolve very slowly indeed.

The human factor: light in the darkness

We are the people our parents warned us about.

Anonymous

Our entire life-cycle, and the course of evolution by natural selection, responds to the diurnal cycle of night and day. It would be easy to think that the existence of night is solely a consequence of the rotation of the Earth and its location relative to the Sun. But it is not. It is a consequence of the expansion of the Universe. If the Universe were not expanding then, wherever we looked into space, our line of sight would end at a star. The result would be like looking into a forest of trees. In a universe that didn't expand, the whole sky would resemble the surface of a star; we would be illuminated by perpetual starlight. What saves us from this everlasting light is the expansion of the Universe. It degrades the intensity of the light from distant stars and galaxies, and it leaves the night sky dark. For roughly half of every day, that darkness silhouettes the Moon and the stars on the vault of heaven. From those silhouettes have flowed all the imaginings, speculations, and impressions that the stars have inspired within us. No civilization is without its stories of the sky and the bodies that shine in the day and the night. Nor are those astronomical impressions of the edge of darkness confined to the distant past, or to cultures still in their infancy. Remember the first pictures of Earth from *Apollo 11*, on its mission to land the first men on the Moon (Plate 7). How striking was that disc of ocean blue, behind the puffs of cotton cloud, set upon a background of total blackness, when seen beside the arid, grey, and lifeless Moon. Those images probably did more than anything else to awaken the collective conscience of humanity to what could be lost through pollution, carelessness, or madness.

The feeling that the Universe is vast and threatening is deeply embedded in the human psyche. The stars appear when the Sun sets, and with them arise perils and uncertainties. Now we know that the stars are too far away to hurt us, but still they can inspire us by their brightness, or depress us by their multitude. The discovery that our Sun is a minor player in the cast of a hundred billion stars that make up our galaxy; that this galaxy is also but one in a population of at least a hundred billion within the visible portion of the Universe alone; all this has given us plenty to be modest about when appraising our place in the scheme of things. It is striking that such a perspective on our part in the cosmic drama could dawn only when we had the technological sophistication to survey and appreciate the structure of the Universe. It comes as a by-product of those same advances in scientific capability that tempt us into a dangerous over-confidence in our powers to control, or ignore, the forces of Nature. The pursuit of pure and applied science is more than simply a matter of balance; 'blue skies' research is more than just a prudent investment in things that might unexpectedly change into profitable industries. It is more than a carrot to appease scientists, or a loss-leader to attract youngsters into the wider ranks of scientists. Maintaining the balance between pure and applicable knowledge about Nature encourages a healthy awareness of the logical depth and astronomical breadth of the structure of the Universe as technology develops; for with it comes a proper humility about our own situation. If pursued in isolation, technology, by its dazzling benefits, threatens to blinker us. Our little successes at manipulating Nature are apt to impress us too much. The fact that a mature picture of one's place in the Universe can emerge only when one has developed those skills, and the accumulated insights that can also pervert it, is a sobering realization. It shows why the advancement of any branch of inquiry naturally engenders new choices and spawns ethical problems. The problems of reconciling our ever-changing scientific view of the world with other things are not a fault of science, or of those other things; nor are they a sign that we have created some grave crisis. These problems are a natural consequence of widening our horizons to an extent that permits us to see ourselves in a new context, which must then be used to judge the very activities that gave rise to it. Any civilization that has developed the technology that would enable it to speak to us across the great deserts of outer space must have encountered the dilemmas that are generated by creating a new scientific picture that includes themselves. If they have spurned or given up the quest for knowledge for its own sake, and have become technicians dedicated merely to their own elaboration and survival, they may lack that clash of views that we call conscience. On our own planet, in recent times, there have been many examples of societies whose rampant technical development has ridden rough-shod over the dignity of individuals and the value of the flora and fauna around us. There is always a tendency for technical possibilities to be overtaken by the worst in our natures; yet, for the most part, the fruits of our pure curiosity, as it seeks out the inner structure of the world, take us by surprise, show us that things are deeper and more rational than we suspected, and reveal that we are more often wrong than right. They have the scope to promote humility and to encourage us to respect the virtues of patience, persistence, and self-correction.

▓▓ The world is not enough: the grand illusion

Nothing is real

The Beatles

Of late, there has been much interest in multiverses. What sorts could there be? And how might their existence help us to understand those life-supporting features of our own universe, that would otherwise appear to be just very fortuitous coincidences? At root, these questions are not ultimately matters of opinion or idle speculation. The underlying Theory of Everything, if it exists, may require many properties of our Universe to have been selected at random, by symmetry-breaking, from a large collection of possibilities and the Universe's vacuum state may be far from unique.

The favoured inflationary cosmological model that has been so impressively supported by the observations of the COBE and WMAP satellites contains many apparent 'coincidences' that allow the Universe to support complexity and life. If we were to consider a 'multiverse' of all possible universes then our observed universe appears special in many ways. Modern quantum physics even provides ways in which these possible universes that make up the multiverse of all possibilities can actually exist.

Once you take seriously that all possible universes can (or do) exist, then a slippery slope opens up before you. It has long been recognized that technical civilizations, only a little more advanced than ourselves, will have the capability to simulate universes in which self-conscious entities can emerge and communicate with one another. They would have computer power that differed from ours by a vast factor. Instead of merely simulating their weather or the formation of galaxies, like we do, they would be able to go further and watch the appearance of stars and planetary systems. Then, having coupled the rules of biochemistry into their astronomical simulations, they would be able to watch the evolution of life and consciousness (all speeded up to occur on whatever time-scale was convenient for them). Just as we watch the life-cycles of fruit flies, they would be able to follow the evolution of life, watch civilizations grow and communicate with each other, argue about whether there existed a Great Programmer in the Sky who created their Universe and who could intervene at will in defiance of the laws of Nature they habitually observed.

Once this capability to simulate universes is achieved, fake universes will proliferate and will soon greatly outnumber the real ones. Thus, Nick Bostrom has argued that a thinking being here and now is more likely to be in a simulated reality than a real one.

Motivated by this alarming conclusion, there have even been suggestions as to how best to conduct ourselves if we have a high probability of being simulated beings in a simulated reality. Robin Hanson suggests that you should act so as to increase the chances of continuing to exist in the simulation or of being resimulated in the future: 'If you might be living in a simulation then all else-equal you should care less about others, live more for today, make your world look more likely to become rich, expect to and try more to participate in pivotal events, be more entertaining and praiseworthy, and keep the famous people around you happier and more interested in you.' In response, Paul Davies has argued that this high probability of living in a simulated reality is a *reductio ad absurdum* for the whole idea that multiverses of all possibilities exist. It would undermine our hopes of acquiring any sure knowledge about the Universe.

The multiverse scenario was suggested by some cosmologists as a way to avoid the conclusion that the Universe was specially designed for life by a Grand Designer. Others saw it as a way to avoid having to say anything more about the problem of fine tuning at all. We see that once conscious observers are allowed to intervene in the universe, rather than being merely lumped into the category of 'observers' who do nothing, that we end up with a scenario in which the gods reappear in unlimited numbers in the guise of the simulators who have power of life and death over the simulated realities that they bring into being. The simulators determine the laws, and can change the laws, that govern their worlds. They can engineer anthropic fine-tunings. They can pull the plug on the simulation at any moment; intervene or distance themselves from their simulation; watch as the simulated creatures argue about whether there is a god who controls or intervenes; work miracles or impose their ethical principles upon the simulated reality. All the time they can avoid having even a twinge of conscience about hurting anyone because their toy reality isn't real, is it? They can even watch their simulated realities grow to a level of sophistication that allows them to simulate higher-order realities of their own.

Faced with these perplexities, do we have any chance of winnowing fake realities from true? What might we expect to see if we made scientific observations from within a simulated reality?

Firstly, the simulators will have been tempted to avoid the complexity of using a consistent set of laws of Nature in their worlds when they can simply patch in 'realistic' effects. When the Disney company makes a film that features the reflection of light from the surface of a lake, it does not use the laws of quantum electrodynamics and optics to compute the light scattering. That would require a stupendous amount of computing power and detail. Instead, the simulation of the light scattering is replaced by plausible rules of thumb that are much briefer than the real thing but give a realistic-looking result—as long as no one looks too closely. There would be an economic and practical imperative for simulated realities to stay that way if they were purely for entertainment. But such limitations to the complexity of the simulation's programming would presumably cause occasional tell-tale problems—and perhaps they would even be visible from within.

Even if the simulators were scrupulous about simulating the laws of Nature, there would be limits to what they could do. Assuming the simulators, or at least the early generations of them, have a very advanced knowledge of the laws of Nature, it's likely they would still have incomplete knowledge of these laws (some philosophers of science would argue this must always be the case). They may know a lot about the physics and programming needed to simulate a universe, but there will be gaps or, worse still, errors in their knowledge of the laws of Nature. They would of course be subtle and far from obvious, otherwise our 'advanced' civilization wouldn't be advanced. These lacunae do not prevent simulations being created and running smoothly for long periods of time. But gradually the little flaws will begin to build up.

Eventually, their effects would snowball and these realities would cease to compute. The only escape is if their creators intervene to patch up the problems one by one as they arise. This is a solution that will be very familiar to the owner of any home computer who receives regular updates in order to protect it against new forms of invasion or repair gaps that its original creators had not foreseen. The creators of a simulation could offer this type of temporary protection, updating the working laws of Nature to include extra things they had learnt since the simulation was initiated.

In this kind of situation, logical contradictions will inevitably arise and the laws in the simulations will appear to break down now and again. The inhabitants of the simulation—especially the simulated scientists—will occasionally be puzzled by the experimental results they obtain. The simulated astronomers might, for instance, make observations that show that their so-called constants of Nature are very slowly changing.

It's likely there could even be sudden glitches in the laws that govern these simulated realities. This is because the simulators would most likely use a technique that has been found effective in all other simulations of complex systems: the use of error-correcting codes to put things back on track.

Take our genetic code, for example. If it were left to its own devices we would not last very long. Errors would accumulate, and death and mutation would quickly follow. We are protected from this by the existence of a mechanism for error correction that identifies and corrects mistakes in genetic coding. Many of our complex computer systems possess the same type of internal 'spell-checker' to guard against error accumulation.

If the simulators used error-correcting computer codes to guard against the fallibility of their simulations as a whole (as well as simulating them on a smaller scale in our genetic code), then every so often a correction would take place to the state or the laws governing the simulation. Mysterious sudden changes would occur that would appear to contravene the very laws of Nature that the simulated scientists were in the habit of observing and predicting.

We might also expect that simulated realities would possess a similar level of maximum computational complexity across the board. The simulated creatures should have a similar complexity to the most complex simulated non-living structures—something that Stephen Wolfram (for quite different reasons, nothing to do with simulated realities) has coined the Principle of Computational Equivalence.

So we conclude that if we live in a simulated reality we should expect occasional sudden glitches, small drifts in the supposed constants and laws of Nature over time, and a dawning realization that the flaws of Nature are as important as the laws of Nature for our understanding of true reality.

3 | Size, life, and landscape

Adaptions to universal features of our world are apt to escape our notice simply because we do not observe anything with which such adaptions stand in contrast.

ROGER N. SHEPARD

A delicate balance: equilibria in the Universe

Uniqueness can be the product of processes that are themselves general to all living matter.

Robert Foley

Size matters. But how does it matter? What determines the sizes of living things, and of the inanimate lumps of celestial matter upon which they have their being? We have uncovered how the ambient conditions in the Universe—the sparsity of matter and the vast star-studded blackness of space—are all consequences of the great age of the Universe. This longevity is essential for the existence of living things: without it there would be neither biology nor biologists. Yet, there is a good deal more to life than simply knowing what is needed of the Universe if it is to be inhabitable. We are going to follow a trail that begins by considering why things have the sizes that they do. This will help us to understand why living things come in a particular range of shapes and sizes. We shall then discover that the ramifications of size are unexpectedly linked to aspects of our past evolution, which influence our present behaviour, and to our penchant for a particular type of aesthetic appreciation.

The Universe around us is filled with a multitude of things. We sit midway between the vastness of intergalactic space and the subatomic microcosm of elementary particles within the atoms of our bodies. To appreciate the diversity of Nature, and to see where we sit in the scheme of things, we should begin by doing a bit of armchair field-work. Send out a team of researchers into the Universe to log the average sizes and masses of all the things to be found there. Seek out everything: from the smallest atoms to the largest clusters of galaxies. Coordinate all this information by plotting a graph of the sizes and masses of the things surveyed. The result looks like Figure 3.1.

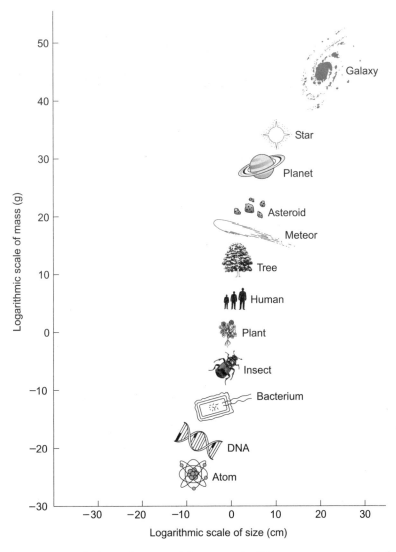

3.1 The masses and sizes of some of the most significant structures found in the Universe.

The picture has an easily discernible pattern. In between the inner space of the subatomic particles and the extent of the entire visible Universe lies a cornucopia beyond imagining: there are clusters of galaxies, lone galaxies of stars, star clusters, shining stars like the Sun, a supporting cast of planets and moons, asteroids and comets; then, smaller still, we find living things, like trees and plants, and a menagerie of animals, fish, birds, and insects, before we reach the microworld of bacteria and cells; delving within them we find molecules great and small; and finally, simple, solitary atoms of hydrogen.

Our cosmic inventory promises to be illuminating. When we take stock of the final picture, a hidden simplicity invites an explanation. One might have expected the things of the world to be scattered all over the picture in a completely haphazard fashion, showing that Nature explores all possibilities. Nothing could be further from the truth. There is a spareness, and a hidden order to things. Large regions of our picture are bare; and Nature's structures form a narrow swathe running diagonally from the top right to the bottom left of the picture. There are evidently regions where there is something, and others where there is nothing. We just need to know why.

First, we need to appreciate what the things around us *are* in the broadest sense. Any structure that we see in the Universe results from a balance between opposing forces of Nature. Since four forces exist in Nature with distinct strengths and ranges, there is scope for a variety of quite different equilibria to exist in which matter is held in the grip of two opposing forces. We are fortunate that the number of natural forces is so small: as a result, Nature is reasonably intelligible to us. If there were thousands of basic forces of Nature, rather than just four, then the number of different structures that could result from any two of them coming into balance would be enormously increased, and the search for simple patterns would be far more difficult—perhaps beyond our ability to deal with.

Not all the structures in our inventory owe their existence to balances between two of Nature's quartet of forces; in some situations the balance is that of a natural force countering motion. Clusters of galaxies and spiral galaxies display the results of a balance between the inward force of gravity, as it pulls stars towards each other, and the rotation of the stars as they orbit around the centre of the cluster.

Individual stars maintain themselves in a balance between the inward squeezing of gravity and the outward pressure of hydrogen gas, or radiation, sustained by the nuclear reactions occurring near the centre of the star. Bodies that are too small for the gravitational squeezing at their centres to attain temperatures of millions of degrees, which are needed to initiate nuclear reactions, can never become stars. Instead, they will remain as the cold bodies that we call 'planets'. Planets are balancing acts in which interatomic forces, which resist any tendency for atoms to overlap, are strong enough to resist the inward crush of gravity.

These simple considerations have revealed to astronomers why planets and stars range in size as they do. Unfortunately, we still do not know whether galaxies and galaxy clusters owe their sizes to a similar balancing principle, or whether they are just residues of haphazard irregularities that came into being with the Universe itself. Galaxies are most probably the largest aggregates of gas that have time to cool and fragment into stars in the time it takes them to contract in size under the influence of gravity. If so, then galaxies will have their sizes determined by the intrinsic strengths of the electromagnetic and gravitational forces of Nature, just as stars do. But clusters of galaxies probably have no similar explanation. During the first million years of the history of the Universe, when it was too hot for atoms and molecules, or stars, to exist, a cosmic sea of radiation was able to smooth out any irregularities in the distribution of matter that were small enough for the radiation waves to traverse. Clusters of galaxies seem to be the smallest irregularities that survived this smoothing process. We would probably have predicted the existence of stars and planets, even if we had never seen either; but I suspect we would not have predicted that the Universe contains beasts like galaxies and galaxy clusters.

A balance occurs between gravitational and atomic forces when matter has a density close to the density of single atoms. Planets, mountains, trees, people, insects, cells, and molecules are all composed of closely packed arrays of atoms. The density of these collections of atoms is therefore similar to the density of a single one of the atoms of which they are made. If two things have the same density it means that the ratio of their mass to their volume is the same. Since volume is proportional to the cube of an average linear measure of size, we see why these solid objects all lie along a line with a slope close to three in the diagram. This is a line of constant density, and that density is atomic density: the density of single atoms. It extends all the way from the simplest atom of hydrogen (which consists of one proton and one electron) right up to the largest solid structures in the Universe. Star clusters, galaxies, and clusters of galaxies are collections of orbiting stars, rather than solid objects; so, they have lower densities, and lie slightly below the line of constant atomic density. We thus see why the things on show in the Universe form such an orderly collection. Despite their superficial diversity, they are linked by a single thread—the similarity of their densities—that issues from the fact that they represent states that can withstand the crushing inward force of gravity.

What about the empty spaces in our diagram? The region set aside for things of small size and large mass is completely empty. Astrophysicists have gradually come to appreciate the reason for this emptiness. It is the realm of the *black hole*. We are familiar with the need to launch projectiles at high speed in order for them to escape the pull of the Earth's gravity. If we throw a stone in the air, then the Earth's gravity brings it back to the ground. Hurl a little faster and it goes higher before returning. But if we launch a rocket fast enough, then it can escape the Earth's pull completely. The bigger a planet, the more material it contains, the greater the pull of its gravity, and the faster you need to launch to escape into outer space. A launch speed of about 11 kilometres per second is required for a rocket to escape the Earth's gravity. As the mass of the planet increases, or its radius decreases, the speed required to escape from the pull of its gravitational field increases. In the nineteenth century an English scientist, John Mitchell, and a French mathematician, Pierre Laplace, both conceived of celestial bodies that were massive enough, and small enough, for the escape of light to be impossible. They would be invisible to all outside observers and could be discerned only by the pull of their gravity. These 'black holes', as they have become known, seem to populate the Universe in great profusion. There is hardly a branch of astrophysics that does not find evidence for their existence, or the need to appeal to their ultra-strong gravitational fields to explain an array of otherwise inexplicable cataclysmic astronomical events. Their existence is the reason for the mysterious lack of massive, small structures in our inventory of the Universe: such things could not be seen. Anything in that part of our picture would be invisible, trapped, incommunicado, within a 'black hole'.

This leaves us with the task of explaining the other empty region of our picture. Again, there is a fundamental limitation upon what Nature allows us to observe, regardless of the sensitivity of our instruments. When we 'see' something, we record a photon of light that bounces off it, directly on to the retina of our eye, or indirectly via the focusing lens of a microscope. If the object we are viewing is large, then the recoil it feels from the rebounding photon is totally negligible. We see the object sharply at a definite location: the photons of light that bounce off a bus into our eye when we cross the road create no ambiguity about the actual position and speed of the bus. But for very small objects, the effect of the recoil can produce a relatively large disruption of what we are trying to measure. To 'see' something, we need to expose it to light of wavelength similar to its size; thus, small objects require small wavelengths of light, which have high frequencies and energies of vibration; these are most able to perturb the system under investigation. This catch-22 situation is expressed by the famous Uncertainty Principle first discovered by Werner Heisenberg. It states that we cannot simultaneously measure the position and the speed of something to ever-increasing accuracy, no matter how perfect our instruments. A measurement of the position of a small object necessarily results in its position being disturbed by the radiation used in the measurement process. It is this consideration of the disruptive effect of the measurement process that distinguishes what is meant when a physicist says that something is 'small' rather than 'large'. 'Small' sounds like a totally relative adjective. 'Smaller than what?' is our response to someone telling us that something is small. But the absolute dividing-line between large things and small things is picked out by asking whether the act of observing them with perfect instruments has a negligible, or a significant, impact on their states.

We have drawn Heisenberg's limit on the diagram (Figure 3.2). It reveals why the last portion of our diagram is empty. Nature is so constructed that we

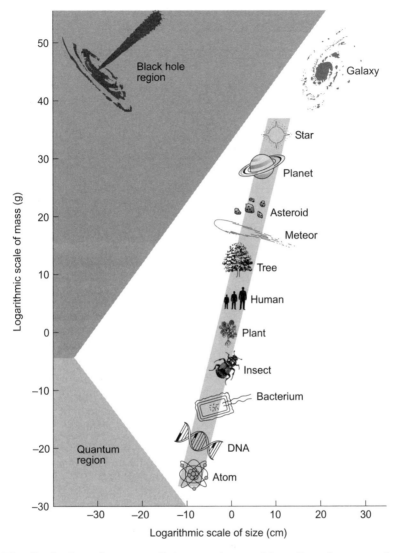

3.2 The distribution of masses and sizes supplemented by a line of constant density equal to that of solid atomic structures, a line bounding the region populated by black holes, and a line bounding the region in which quantum uncertainty renders objects unobservable.

cannot 'see' any objects in the bottom triangle of the picture without disrupting them by the very act of observation. There is no way of measuring things that is discrete and gentle enough to allow us to probe the 'uncertain' region. True states in the quantum region would always be seen to lie on, or above, Heisenberg's line because of the measurement process.

An important lesson from these simple considerations is that the sizes of things in the Universe are not random. They lie in the ranges they occupy because they are manifestations of balances between opposing forces of Nature. Their locations are determined by the inflexible values of the constants of Nature that express the strengths of those forces. This is the explanation for the things that are seen. As for the things that are not seen, the existence of black holes and the Heisenberg Uncertainty Principle exclude a vast range of mass-size combinations from the view of any observers. The cosmic inventory is neither the result of some process of natural selection, nor of an attrition of the set of all possibilities; and it is not merely haphazard—it is at once a matter of balance and censorship.

▓▓ Of mice and men: life on Earth

'I should like to be a *little* larger, Sir, if you wouldn't mind,' said Alice: 'three inches is such a wretched height to be.'
'It is a very good height indeed!' said the Caterpillar.

Lewis Carroll, *Alice in Wonderland*

Let us move a little closer to home. As we move up the line of constant density in Figure 3.2, the strength of gravity at the surface of each object increases linearly with its size, and it is harder to escape from its surface. This simple difference between large bodies and small ones has the most far-reaching consequences. To understand why, we need only compare the Earth and Moon.

The Earth and the Moon are totally different. The Earth is surrounded by a biosphere of stupendous complexity. The Moon is arid and dead. The reason: the Earth has an atmosphere of gases, like nitrogen and oxygen, that promote intricate chemical and biological processes on its surface, while the Moon has none. The Moon is too small for the pull of its gravity to retain a gaseous atmosphere. If the air in this room were teleported to the Moon, the gas molecules would be moving fast enough to escape the pull of the Moon's gravity and they would rapidly disperse into space. Hence, only planets bigger than a critical size will possess atmospheres and offer the possibility of biochemistry. Yet, life-supporting planets must not be too big. Living things are made out of intricately organized collections of atoms and molecules, held together by interatomic and molecular bonds. These bonds can be broken by excessive pressure or temperature. When they are, matter changes its properties, often irreversibly. Fry an egg, and you see what happens to the protein molecules in the egg-white when the temperature gets too high. Suddenly, they lose their fluid mobility and become rigid as the proteins in the egg-white solidify. When this change of state occurs, we say that the egg is 'cooked'. 'Cooking' is simply the attainment of the temperature at which intermolecular bonds are changed—or 'denaturized' as chemists say. (Others in my household assure me that there is more to it than that.) Likewise, delicate molecular bonds will be ruptured if they are crushed, or stretched, by strong forces. Living things composed of huge numbers of atoms and molecules, held together by a lattice-work of interatomic bonds, are therefore living rather dangerously. Put them where it is too hot, and their complex molecular bonds will be cooked into immobility. Put them on a planet that is too big, and they will be crushed by the overwhelming strength of gravity at its surface.

Habitable planets must therefore be neither too big, nor too small. Only the in-between worlds, like the Earth, combine the possibility of retaining an atmosphere with surface conditions that are moderate enough to permit the presence of complex molecular architectures. Even on Earth, the force of gravity plays a key role in limiting the scope of living and inanimate things. The need for bonds between atoms to hold molecules together is the reason why terrestrial mountains cannot be much higher than Mt Everest. As a mountain gets higher, and heavier, so the pressure on its base increases. If it were too high, then the bonds between atoms would start to break, and the mountain would just sink into the Earth's crust until the pressure at the base was reduced sufficiently for the material there to solidify. Living things—trees and birds, land-going animals, and sea creatures—are also strongly constrained in size by the forces of Nature. A tree that grew too large would suffer unacceptable pressures on its base, and break. In practice, it is the susceptibility to breaking when bent by the wind that snaps the tree and limits its maximum height. Trees cannot grow without limit, because their strength does not keep pace with their size as they grow.

We thus begin to see how the living world around us is shaped by the forces of Nature through a long chain of connections. The relative strengths of gravity and the atomic forces of electromagnetic origin closely determine the sizes of planets with atmospheres. The size of structures that can stand up, or move safely, on the surfaces of those planets is also limited by the destructive crushing of their atomic bonds by gravity. There are limits to the sizes of living things because strength cannot keep pace with size and weight as they grow. This failure of strength to match increasing weight and volume is evident if we watch creatures of different sizes. An ant can carry a load that is ten times greater than its own body weight (Figure 3.3). A small dog can easily carry another dog on its back. A child can carry another child, piggy-back, without too much trouble; but an adult has much more difficulty; and no horse is strong enough to carry another horse on its back. As you get bigger, the stress on your bones gets greater; they need to be larger and thicker to stand the strain. A graphic example of this disparity between strength and size is displayed by a homely example. A small kitten's tail will stand bolt upright like a spike, because the kitten is strong

3.3 An ant is strong enough to carry a load many times greater than its own body weight.

enough to support its little tail. But look at its bigger mother. Her longer tail loops over, because she is not strong enough to hold it upright.

The strength of things is determined not by their weight or by their volume, but by the area of a slice through them. When bones break they do so along a sheet of bone. It is the number of bonds that need to be broken in that sheet that determines its strength. The mass of the body is determined by its total extent; its resistance to breakage is determined by the structure of the areas where it is subjected to stress. Visually, we judge the body-builder's strength by the size of a cross-section through his biceps.* The strength of a rod is determined by its thickness at the place where we try to snap it, not by its overall length or total volume. Nevertheless, for most things, there is a simple relationship between the average area of a slice through them and their total volume or mass. An area is proportional to the square of some average measure of the length of a thing,

* An examination of the world weightlifting records shows this dependence, with the record weight lifted proportional to the two-thirds power of the body weight of the lifter.

whereas its volume (and hence its mass, if the density is constant) is always proportional to the cube of that length. If we pick that length to be an average diameter of the object in question, then as it grows in size its strength will increase as the square of its size, but its weight will grow in proportion to its volume, and so to the *cube* of its size. Thus, as it gets bigger, it is less and less able to support its own weight. There is a maximum size, after which it simply breaks.

One of the largest dinosaurs was *Apatosaurus* (previously called *Bronto-saurus*). At 85 tons, it was pretty close to the size limit for a land-going animal. It had little room for error if it were to stumble, or transfer too much of its weight on to one leg. (For comparison, the largest land-going creature today, the African elephant, weighs only about 7 tons.) Walking up the slightest incline would have been extremely taxing for a large dinosaur, because it would then have to lift a component of its body weight against the downward force of gravity. The heavier you are, the slower you are able to move uphill. Dinosaurs alleviated the pressures on their base by spreading their load over their widely spaced legs. This aids stability; but none the less, if they fell, they would prob-ably break their bones. Adult humans have a much shorter distance to fall if they stumble, but sometimes still break bones. Children fall shorter distances, and don't often break bones, despite constantly taking tumbles. (The fact that young bones are softer, and less brittle, than old ones also helps.) An adult may hit the floor with an energy of motion more than six times greater than that of an infant when they both stumble and fall. If adults were twice as big as they are, then walking upright would be a very dangerous business—rather like walking on stilts.

One way of overcoming the failure of strength to keep pace with volume is to exploit buoyancy. When any object is placed in a liquid medium, like water, it feels a buoyancy force pushing it upwards, equal to the weight of liquid it displaces. As a result, the stress on its base is alleviated. It is no accident that the largest blue whales (at 130 tons) are enormously bigger than the biggest land-going dinosaurs ever were, or ever could have been.

Water can also support you if you are small. Place a paper-clip on the undisturbed surface of some water, and it floats, supported by a force of 'surface tension' that arises at the interface of its surface with the water. If you add detergent to the water, then this force is reduced and the clip will eventually sink. Very small creatures, like pond-skaters, can use surface tension to support their weight so long as they have their legs spread out over a few square millimetres of surface. Again, the surface force increases more slowly with increasing size than does the weight of a creature; there is thus a maximum size, and weight, that can be supported in this way. It works only if you are very small. Humans would need legs spread over about 7 kilometres in order to walk on water, like Hilaire Belloc's water beetle:

The water beetle here shall teach
A sermon far beyond your reach:
He flabbergasts the Human Race
By gliding on the water's face
With ease, celerity, and grace;
But if he ever stopped to think
Of how he did it, he would sink.

If you want to fly, then it also pays to be small. Your wings must generate enough lift to overcome the pull of gravity. As you get bigger, the power required to support your weight grows faster than the power your muscles can exert. Consequently, there is a maximum size for a flying creature. The largest birds that can hover in still air for long periods are humming-birds; they vary between about 2 and 20 grams in weight. In fact, they can even take off vertically. Of course, there are far bigger birds. At the top of the tree are the largest Kori bustards, weighing about 12 kilograms. But they stay aloft by soaring, riding wind-currents, or thermal draughts. Likewise, when you see a kestrel hovering over a point on the ground, it is not hovering; it is flying against the wind—just strongly enough to ensure that it remains stationary relative to the ground. It is not strong enough to support its own weight by hovering in still air.

These examples reveal something of the ongoing terrestrial battle between strength and weight, which pits the force of gravity against the intermolecular forces of electromagnetic origin. It was these same forces that first determined the inevitable sizes of habitable planets with atmospheres, the strength of gravity at their surfaces, and, hence, the sizes of complex living things that can exist on their surfaces. Our size is not an accident. It is, within quite narrow bounds, imposed by the invariant strengths of the forces of Nature. But the consequences of our size for our development, our culture, and our abilities are deep and wide. They shed light upon how we have outstripped other living things in controlling natural resources.

The battle between strength and size is displayed by a simpler struggle: that waged between volume and the surface area enclosing it. Watch a rolling snow-ball as it accretes snow and grows bigger. Its radius increases; so both its volume and surface area grow as well. But, whereas its volume grows according to the cube of its radius, its surface area increases only in proportion to the square of the radius: its surface area cannot keep pace with the growth in its volume. This losing battle that surface area fights with volume as size increases imposes many vital constraints upon the sizes of living things. As your volume increases with growth, so your heat-generating organs increase in volume and energy output. But your ability to keep cool depends upon how much heat can escape from your exposed surface. Small creatures have a relatively large surface area for their volume; large creatures possess a relatively small one. In cold climates small creatures will therefore be at a disadvantage, unable to generate enough heat from eating to keep warm. This is why babies need to be wrapped up so much more thoroughly than adults in cold conditions. Large animals, by contrast, will be at an advantage in the cold. Hence we find large animals—polar bears, rather than small mice—at the Poles, and the average size of birds increases between the Equator and the Poles. The smallest shrews are about as small as animals can be without being thermodynamically at great risk in an environment that occasionally cools to a few degrees below their body temperatures. Small animals can combat this risk of exposure by huddling together for warmth. They share their body heat, and reduce their exposed surfaces by mimicking the geometry of a larger creature. Small animals can also adopt other strategies for reducing their heat losses: growing fur, for example; or even, like some creatures we know, by wearing other creatures' fur to provide insulation when required.

Their surface area also determines how fast materials will burn, because the exposed surface is where oxygen is consumed to sustain the flames. Small objects have relatively more surface area than large ones; and the further they are from a spherical shape, the greater will be the surface they expose. This is why wood shavings, just a fraction of a centimetre in size, burn so much faster than a round log 10 centimetres in diameter. When we want to start a fire with pieces of paper, we screw them up rather than lay them flat, so as to increase the amount of exposed paper surface.

We grow used to discovering that some naturally occurring structure is the best possible by some engineering criterion. But, as we stressed in the last chapter, it is wrong to believe that all Nature's solutions to the problems posed by the environment are optimal. They may not need to be. Nor might the range of variations available for natural selection include the optimal case, because of other constraints on what can happen, or simply because of bad luck. An interesting example of this sort is the honeycomb that bees construct. This involves the problem of optimizing the surface area enclosing a particular volume. The early Greeks had speculated that there must be some hidden principle of optimality that accounted for the bees' symmetrical honeycombs; by the seventeenth century, the problem was recognized as the search for a pattern that minimized the amount of wax that was needed to create a network of cells. The honeycomb is a network of prism-shaped cells, whose tops are hexagons with equal sides, but whose bases are each composed of three diamond-shaped planes (rhombi), which are joined to the sides of the hexagon (see Figure 3.4). The zig-zagging of the bottom surface of the honeycomb certainly makes for a more economical use of resources than a flat surface; but is it the best possible shape?

In 1964, the Hungarian mathematician Fejes Tóth posed the 'honeycomb problem' as the determination of the cell shape to be used when constructing a honeycomb of a given width, enclosing a given volume, so that the cells present the

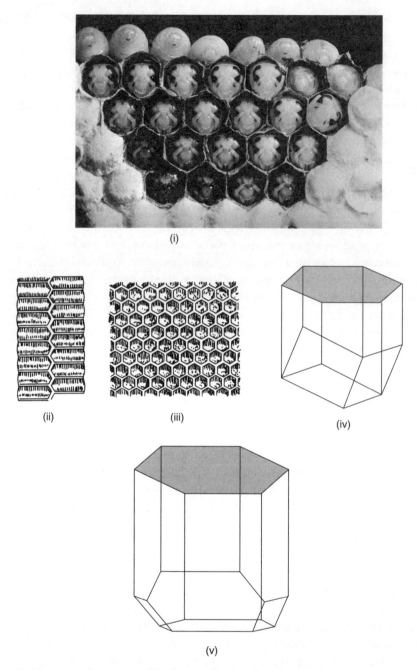

3.4 (i) A honeycomb network; (ii) a longitudinal section through the honeycomb; (iii) a cross-section through the honeycomb; (iv) an individual cell of the honeycomb; (v) the cell shape discovered by Fejes Tóth, which is more economical on materials than that used by the bees.

smallest possible surface area. The problem is still unsolved. So far, no one has discovered what the most economical cell shape is; but what is known is that it cannot be the one that the bees use. Tóth found a base pattern, using two hexagons and two diamonds, that did better than the bees; but it manages to economize on surface only by less than one per cent of the area of the hexagonal top of each cell. The bees could therefore do better, although not, it appears, by very much.

The jagged edge: living fractals

Round the rugged rocks the ragged rascals ran.

Nursery rhyme

The study of surfaces that enclose a particular volume of space by an anomalously large surface area has become very fashionable among mathematicians and computer-graphics *aficionados*. Such surfaces are examples of what the French-American mathematician Benoit Mandelbrot has called 'fractals'. Fractals can be constructed by copying a basic pattern over and over again, each time on a smaller and smaller scale. Spectacular fractal images can be found all around us, on posters and magazine covers—there have even been exhibitions of computer-generated fractal 'art' at major international galleries—but there are more serious applications as well. The intricacy of fractal designs offers a way of boosting the capacity of computer memories, and minimizing the effects of vibrational disturbances in mechanical structures.

We see that Nature uses fractals everywhere: in the branching of trees and the shaping of leaves, flowers, and vegetables. Take a look at the head of a cauliflower, or a sprig of broccoli, and you can see how the same pattern is repeated over and over again on different scales. What an economical plan for the development of complexity! Another reason for the ubiquity of fractal designs in Nature is that they offer a general recipe for escaping the strait-jacket on design that is imposed by the simple relation between volume and surface area that we find in regular objects, like our rolling snowball. By allowing the surface of a ball to become intricately crenellated, its exposed surface can be greatly increased over that needed to enclose the same volume smoothly. Examples of fractal surface enlargement abound (Figure 3.5). Our lungs display a branching fractal network of tubes that maximize the absorption of oxygen through their surfaces. Sponges have far more surface than a solid ball of the same volume in order to increase their surface exposure to the organisms they ingest. When we step out of the shower, we dry ourselves with a towel that exhibits a surface of tiny knots. They enlarge the area of towel that makes contact with the body, and so increase the uptake of moisture by the towel.

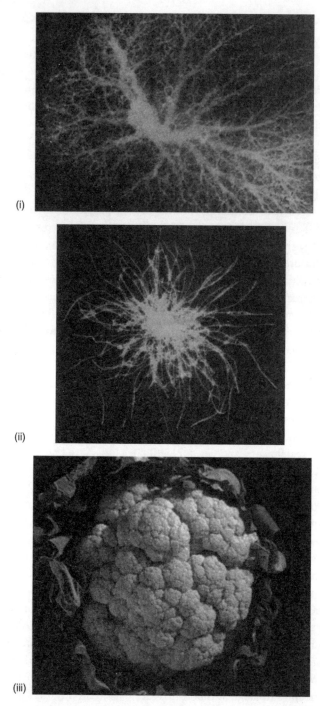

(i)

(ii)

(iii)

3.5 Some examples of fractal patterns appearing in Nature: (i) the human lung; (ii) a piece of eider down; (iii) the head of a cauliflower.

Wherever there is a need to expose as large a surface as possible, but there is a restriction on the total volume of material available, or a penalty to be incurred by increasing weight, then fractals are selected by the evolutionary process. Fractal structures are also good at damping down vibrations. For example, if one were to make a drum with a fractal-shaped edge, then a beat of this drum would be quickly dulled. Fractal shapes may therefore be extremely robust in situations like those of trees in the wind, panting lungs, or pounding hearts, where it is necessary to withstand a large amount of associated vibration.

The more we scrutinize the structure of Nature, the more fractals we find. Indeed, their ubiquity in the natural world of which we are a part is one reason why we find them so comfortingly attractive. They are a form of computer art that has captured the essential program—self-similar reproduction of the same pattern in different sizes—that living systems have employed to establish their own distinct niches throughout evolutionary history. It could be argued that works of computer-generated fractal art fall short of being art in a form that we find interesting enough to examine and *re-examine* precisely because they are purely self-similar. Only when there is occasional departure from exact self-similar reproduction does the image become artistically interesting, rather than simply symmetrically pleasing. Sometimes the less demanding pleasure of symmetry is what we require. We would not enjoy a wallpaper in the sitting room that provoked the brain to engage in endless repeated bouts of analysis and interpretation whenever we rested our eyes upon it. We prefer more challenging pieces of composition to be framed within boundaries that signal their nature and alert the mind to the interpretational challenge that is about to be set. At the end of this chapter, we shall have a lot more to say about the issue of whether 'fractal art' is really art.

The tortuous textures of fractal surfaces draw our attention to the question of symmetry and shape. Living things are strikingly symmetrical. Whereas inanimate objects rarely display perfect symmetry, animals often possess bilateral symmetry—externally, at least. This is an improbable state of affairs; it witnesses to delicate engineering, and is achieved at high genetic cost. Symmetry is absent in the up–down direction because bodies are adapted to accommodate the variation of the force of gravity with height, and to the need to remain stable in the face of small disturbances that would otherwise push them over. A low centre of gravity on a broad base is here most advantageous and leads to a tapering of the body with distance from the ground. Deviations from a symmetrical bodily form invariably signal some injury or genetic impairment. Some of the worst consequences of disease arise from the loss of our delicate body symmetry. Many of our evaluations of physical beauty focus upon the symmetries of the human facial and bodily form; plastic surgeons receive large sums for restoring or enhancing it. Among lower animals, the perfection of bodily form is an important indicator in the selection of a mate and in distinguishing fellow-members of the same species from predators.

The biggest pay-off from a bilaterally symmetric body comes when you want to move around. In situations where survival is enhanced by an ability to move in a predetermined manner, symmetrical organisms have an advantage. The imbalances created by asymmetries make straight-line motion difficult to engineer; symmetry ensures that linear movement arises in response to thrusting limbs. The benefits of symmetry are even greater if that motion has to take place in water or air. The avoidance of head–tail symmetry witnesses to the higher cost of engineering an arrangement that allows forward and backward motion with equal ease, rather than endowing head–tail asymmetry and an ability to turn round.

The symmetry of animal bodies also has implications for the brain and the senses. The responses of a nervous system need more complicated processing if they have to create a mental body-map by monitoring the surface activity of an asymmetrical periphery. Yet, in contrast, when one examines the layout of the brain itself, it is highly asymmetrical. One side of the brain generally governs the opposite side of the body, and there is a division of cognitive activities between the two sides of the brain. Here, one sees a situation where symmetry would be costly and inappropriate. If all activities were controlled by a symmetrical distribution of neural networks, situated in both hemispheres of the brain, then duplication of activity and waste of resources would be occurring. Such a duplication would not succeed in competition with systems that avoided it, unless there was such a high failure rate of brain function in one hemisphere that it paid to install a back-up system in the other. Such a situation would not evolve. The asymmetry of brain structure reflects the optimality gained by having some circuits close to others. Much of the brain's control is exercised over sequences of operations that need to be meticulously coordinated, and the asymmetrical layout of the brain's programming reflects the need to associate the control of linked body movements and senses. Because the brain governs motion, but does not need to move itself, it can be programmed asymmetrically.

■ Bilateral agreements: appreciating curves

> 'A figure with curves always offers a lot of interesting angles.'
>
> Mae West

Our liking for lateral symmetry seems to owe something to the evolutionary advantages once offered to those with a sensitivity for it. It influences our evaluation of facial beauty in other people and defines cultural norms to which many people aspire in terms of bodily appearance. But there are also particular traditional art-forms that have exploited our satisfaction with lateral symmetry in its purest form. The most impressive is in the production of vases. The vase form is an attractive one for study because it is a relatively simple matter. If we consider vases that are rotationally symmetric, as in the Chinese tradition, then we need consider just a two-dimensional profile in order to highlight what it is about the shape that we find appealing. It is laterally symmetric, but there are an infinite number of variations that can be added to the visual contour of the vase whilst respecting its perfect left–right symmetry. There is a long tradition of seeing aspects of the human form in the shapes of vases, as if they were works of sculpture. The terms used to describe parts of a vase, the 'lip', the 'neck', or the 'foot', bear witness to this.

The American mathematician George Birkhoff maintained a lifetime interest in aesthetics and tried to develop simple ways of quantifying aesthetic appeal in order to capture what impressed us most about artworks and to see what happens if we create new works that deliberately maximise these 'aesthetic measures', as he called them. In 1933 he wrote a wide-ranging book about ways in which some aspects of aesthetic appreciation and its search for 'unity in diversity' could be captured by simple measures. Generally, within some class of similar artistic creations, Birkhoff's measures have the quotient form

Aesthetic Measure = Order/Complexity

This is an attempt to quantify some simple intuitions. We like order and symmetry and so increased order increases the measure, but if complexity rises as well then our appreciation is reduced. Of course, this is an oversimplified approach to both order and complexity, but the real question is how to define the order and the complexity in this formula.

In the case of vase profiles viewed in projection we are faced with evaluating a two-dimensional shape like that in Figure 3.6 if we ignore colouring and texture. There is always lateral symmetry and two curvilinear sides with two circular or elliptical ends (we will ignore the possibility of lids for simplicity).

Our aesthetic appreciation of the vase profile is influenced by several simple geometrical features: places where the contour ends (the lip and base of the vase), places where the tangent to the contour is vertical, places where the tangent changes direction abruptly (the corner points), and points of inflexion in the tangent direction where curvature changes. These special points on the contour create our aesthetic impression of its symmetrical form, see Figure 3.7.

Birkhoff chose to define the *complexity*, C, of the vase form to be the number of special points where the tangent to its contour is vertical, has inflexions, corners, or end points. By inspecting the possible outline in Figure 3.7, we see that the complexity must lie between 6 and 20. Gauging the *order*, O, of the profile is a little more complicated. Birkhoff defined it to be the sum of four factors:

3.6 A typical classical vase profile.

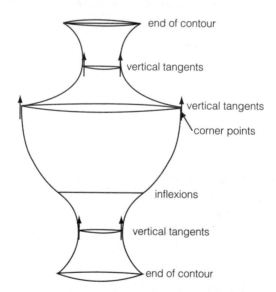

3.7 A vase contour with critical visual cueing points indicated. Not all of these points need occur, but those that do will strongly influence the aesthetic impact.

a.   **H**: the number of horizontal distance relations that are in a ratio of one to one or two to one. This is always less than four.

b.   **V**: the number of independent vertical distance relations that are in the ratio one to one or one to two. This is always less than four.

c. **HV**: the number of independent interrelations between vertical and horizontal distances that are in the ratio one to one or one to two. This is always less than two.

d. **T**: The number of independent perpendicular and parallel relations between tangents plus the number of vertical tangents at end points and inflexions and the number of characteristic tangents through an adjacent centre. This is always less than four.

Therefore we see that the order must always be less than 14. The Aesthetic Measure, M, is determined by dividing the *order* by the *complexity*, so M = O/C, and it can never be greater than $14/6 = 2.333$.

Four classic Chinese Ming, Sung, and T'ang vases were analysed in this way and the numbers that emerge are shown alongside their profiles in Figure 3.8. Their M-values are 0.8, 0.625, and two of 0.583.

Birkhoff then created some experimental classical vase forms of his own with a view to maximizing the ratio of O/C. They are shown in Figure 3.9. They have M-values of 1 and 1.08 that are much larger than the real vases shown in Figure 3.8.

These examples are not meant to try to characterize uniquely the visual appeal of vases; that was far from being Birkhoff's aim and even further from being his conclusion. Rather, it shows how specifically geometrical considerations can capture some of the things that we like about curves and profiles. How those appealing and eye-catching elements are actually weighted in any formula is an entirely subjective matter.

Birkhoff used this approach to evaluate the visual appeal of simpler patterns, in which only straight lines appear, that are used in tiling ornaments. In this situation the measures of complexity and order are much more complicated and there are many more types of shapes that can be created. The more constrained the problem, the more meaningful is any comparative numerical measure of aesthetic impact, although none can cater entirely for individual taste.

Fractal expressionism: the strange case of Jack the Dripper

'The love of complexity without reductionism makes art; the love of complexity with reductionism makes science'

E. O. Wilson

We have seen how fractals populate the natural world around us. They are an ubiquitous solution to the problem of maximizing the amount of surface area a body possesses without increasing its volume and weight at the same time. Trees compete for light and moisture and so the greater the surface of leaves that they have in contact with the air the better they will fare. The fractal recipe for

Ming
C = 10
H = 3, V = 2, HV = 1
T = 2, O = 8
M = 0.8

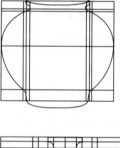

Sung
C = 8
H = V, HV = 1
T = 2, O = 5
M = 0.625

T'ang
C = 12
H = 3, V = 2, HV = 1
T = 1, O = 7
M = 0.583

Sung
C = 12
H = 3, V = 2, HV = 1
T = 1, O = 7
M = 0.583

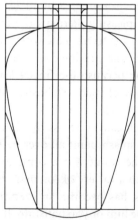

3.8 The Order, O, Complexity, C, and Birkhoff's Aesthetic Measure, M = O/C, of four classic Chinese vase forms.

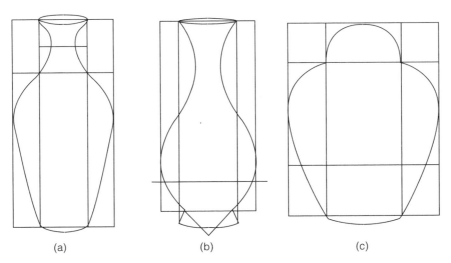

3.9 Three artificially generated vase forms with high Aesthetic Measures of (a) M = 1.0, (b) M = 1.0, and (c) M = 1.08. These all exceed the values of Aesthetic Measure found for the actual vase profiles shown in Figure 3.8.

creating a lot of surface is to copy the blueprint again and again always with a smaller (or larger) size. The branching of a tree follows this recipe as you follow it from the trunk down the succession of branches to the tips of the smallest twigs. When the branching angle is small we end up with a tall tightly branching tree like a fir or a poplar, but when as the branching angle becomes larger we end up with a spreading tree like an oak or a plane. As we have just discussed, our liking for mathematically-generated fractals owes much to our biophilia for the natural environment around us. Trees have 'interesting' shapes that are aesthetically pleasing. Perhaps the fractal basis for these shapes can be artificially captured so as to resonate with the same innate aesthetic sense?

There is one striking case where this appears to have happened. It is especially interesting because the fractal aspect was only detected more than fifty years after the work had established itself as artistically compelling. The work in question is Jackson Pollock's abstract expressionism. In 1947 Pollock unrolled a huge blank canvas across the floor of his barnlike country studio and began to drip and splash paint all over it. The paint was poured in a continuous stream so that the result was a complex network of continuous lines weaving crazy patterns in a variety of colours.

Pollock has become an exemplar of abstract art to many non-artists. You either love him or you hate him. If you hate him, then you tend to think that your three-year-old child can do what he did and you can't understand why his works are valued up to $40 million dollars today. Indeed, no other modern artist attracts such high prices. Then again, if you look at many of Pollock's works in the gallery alongside other works of abstract expressionism there is something different about Pollock that is not easy to capture. And it is certainly different to your child's work. There is a similarity from work to work in the face of obvious difference in pattern. It is abstract, it is seemingly random, yet there is an ordered quality about the work. There is colour but its role seems secondary. Between 1947 and 1952, the period of his greatest work, Pollock often referred to his paintings as 'organic' in form, suggesting an affinity with natural patterns and complexities. He saw his paintings as specially created environments with no centre of focus, like vast landscapes devoid of symbols and signs.

The mystery and appeal of Pollock's work has been significantly unravelled by Richard Taylor's investigations into its fractal structure. A lot of film exists showing Pollock at work on his canvases and his technique displays an order amid apparent chaos. He laid down pattern in layers, eventually abandoning contact with the canvas completely, dripping paint from syringes and trowels, or running it down long sticks on to the canvas in sweeping movements.

Gradually he mastered more sophisticated aspects of the method that relied upon using paint of the right runniness. After a first ground-setting pattern, Pollock would build up the detail and intricacy, working on smaller scales. Long periods of reflection would occur during which he would work on other paintings before returning to develop older ones. He hated the idea of finishing, especially signing, his works. It was as if that signalled the end of their lives. Canvases were sometimes stretched or subjected to bombardments with small fragments. To the observer it all looked as random as could be.

Taylor, who was both a student of art and of physics, decided to apply some simple pattern recognition analysis to Pollock's work—fortunately the originals were not needed—to discover whether there was some pattern being imposed by Pollock's intuition that was being missed by the traditional art critics. The outcome was rather surprising. He discovered that Pollock's paintings are almost perfect fractals with a well-defined range of fractality that developed during his career. The fractal structure means that statistically Pollock's paintings look the same whether you magnify or reduce them in size. We say that their patterns are 'scale invariant'. They have no defining size. You should not be able to tell, just by looking, whether you are viewing an actual size Pollock or a reduced image. If you buy one, cut it in half and sell one of them!

By overlaying a digital scan of a Pollock canvas with a series of increasingly fine grained meshes of squares—like the old-style squared paper you used to use for arithmetic at school—we can determine how many squares have paint in them and how that number changes as the fineness of the mesh increases. This variation is called the fractal dimension. It will vary between the values 1 and 2. If the pattern is very simple—one straight line—then the number of mesh squares that have paint in them will decrease like L^{-1} as we increase the mesh size, L. On the other hand, a very convoluted and complicated line that covered almost all the surface would exhibit a change in the number of covered squares that fell as L^{-2} as we increased the mesh size. The number D appearing in the power L^{-D} is the fractal dimension of the pattern. It tells us how much information is contained in the pattern. A very simple straight-line pattern has $D = 1$ and is line-like. However, the complicated pattern, although still generated by a line, covers a two-dimensional area as if it is two-dimensional. Unlike the ordinary geometrical measure of dimension, this one has the feature that it can be fractional. In between $D = 1$ and $D = 2$ there is a whole range of patterns of intermediate complexity with fractional (i.e. 'fractal') dimension. Some examples are shown in Figure 3.10.

The results of the investigations by Taylor and his collaborators are striking. All of the 23 works by Pollock that they analysed followed the L^{-D} square-counting rule characteristic of a fractal. These paintings covered the whole period of his working life and included works covering a wide range of sizes. When looked at closely, they found that there are two processes going on in Pollock's work. Over the largest dimensions of the canvas films of him at work reveal that he is effectively throwing paint, using his body movements to cover the canvas from scales of a few centimetres up to nearly two metres. But when we look at smaller scales, from millimetres up to a few centimetres, we see the effects of the dripping process. Accordingly, down on small dimensions the fractal index clusters around 1.5 to 1.7, while up on the larger scales it is around 1.95, reflecting the changeover between the two behaviours (see plates 22 (a) & (b) and 23 (a) & (b)).

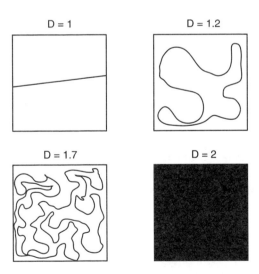

3.10 Patterns with low, medium, and high fractal dimension.

These studies reveal that Jackson Pollock had intuitively sensed the characteristics of fractals long before Benoit Mandelbrot attracted people's attention to these mathematical objects and gave them their enduring name. Fractals have patterns on all scales and the eye is not drawn to a particular dominant scale of statistical pattern. Pollock was extremely perceptive in his identification of this statistical feature by eye, from experience. If you do a similar analysis of the paint all over the floor of Pollock's studio—the drips that missed the canvas—then it is comforting to discover that they do not have a fractal pattern. Pollock's patterns are not accidental.

Pollock's works are fractal to a high degree of accuracy, so much so that Taylor and his colleagues have been involved in authenticating a work of Pollock of unknown provenance and discounting others as true Pollocks because of the presence or absence of the distinctive fractal patterns and transition from small-scale to large-scale structure (see plates 22 (a) & (b) and 23 (a) & (b)). So far, ten unattributed drip paintings from US collections have been analysed in the hope (of their owners) that they might be true Pollocks. Alas, none have the Pollock fractal signatures on small and large scales. Superficially, these works look like the real thing. They have a colouring and rough style that could be mistaken for Pollock's. But only a fractal analysis homes in on the key Pollock ingredient that reveals their true pedigree.

▓▓ War and peace: size and culture

Joe Gillis: 'You used to be in pictures.
  You used to be big.'
Norma Desmond: 'I am big. It's the pictures that got small.'

Charles Brackett, *Sunset Boulevard* (film, 1950)

There have been many attempts to create fantasies peopled by giants or midgets: Bunyan's Giant Despair and Gulliver's travels among tiny Lilliputians and gigantic Brobdingnagians have entertained readers for centuries (Figure 3.6). In modern times, they have been joined by armies of giant insects and super-heroes that walk the pages of the comic books that launched a thousand awful B-movies in the 1950s. Sadly, these inflated beings are impossibilities of structural engineering. The failure of strength to keep pace with volume and weight means that if you simply scale up the whole human body it will be liable to break when it reaches about 600 pounds in weight. To support a greater weight in a state of motion it would need to be redesigned—with shorter and wider bones, wider foot-placement, and quite different internal organs—in order to provide the extra power needed to move the monster around.

Clearly, our size—and we are the largest creature that walks on two legs—has

3.11 Gulliver in Lilliput, by C. E. Brock, 1894.

influenced many aspects of our technological and social development in ways that are both good and bad. It is even possible to argue that our size has been the most important enabling factor in our development of complex technology, and the many social, cultural, and artistic activities that flow from it. We are large enough, and hence strong enough, to wield tools that can transfer enough energy to split rocks and deform metals. This is a consequence of our large size. As the scale of a living creature diminishes, although it may be relatively stronger in terms of the number of its own body weights that it can lift, its absolute strength decreases.

The strength of rocks and metals is fixed by the strength of the electro-magnetic forces of Nature, and by the masses of protons and electrons. When a creature's size falls below a critical level, it will not be able to break the molecular bonds in solid materials. Our own size has enabled us to chisel and excavate rock faces, to split wood, and to hammer metal. By these means, our large size has enabled us to exploit the environment in ways that are closed to smaller organisms. Of course, with the passage of time, we have developed artificial aids for cutting and shaping hard materials, so that we are no longer limited by the strength of our bodies. But these sophisticated secondary abilities cannot arise without the earlier use of manual force. Clearly, the evolution of our unique bipedal stance was of importance in permitting the development of our manual dexterity. It also played an important role in assisting our mobility. Not only does it give us more agility, but early in the evolutionary process it allowed creatures to stay cool in hot climates more efficiently than if they had walked on four legs. During daylight hours in the tropics, less surface is exposed to absorb radiation and, by being farther from the ground, the head (and hence the brain within it) is kept significantly cooler suspended on two legs than it is on four.

But we can also use those tools as weapons that transfer enough momentum to kill other living things for food, for protection, or for no reason at all. Again, our size happens to be appropriate for killing small animals with simple weapons, like rocks or crude mallets. Our ability to deal lethal blows on fellow humans is also a consequence of our size. The scope and consequences of violent action, leading ultimately to warfare, flow from the particular level of strength that goes with our size. If we were only a quarter of our actual size our history would be very different indeed.

Our ability to use fire is also connected with our size. There is a smallest possible flame that burns in air, because the surface area surrounding a volume of burning material determines the influx of oxygen that can sustain combustion. As the burning volume decreases, the surface area falls faster, and the fire is increasingly starved of oxygen. Eventually, a limit is reached, at about half a centimetre, below which the flame cannot be sustained. The initiation of burning requires a temperature of a few hundred degrees Celsius to be achieved. If the temperature falls below this value the flame will die. The size of the flame must therefore be large enough to maintain the burning temperature in the presence of cold air rushing in from outside. If the flame is too small, then those air currents will cool it enough to extinguish it. Indeed, we recognize that any flame is vulnerable to extinction by strong draughts of wind like this, and so we shield our first attempts at lighting the camp-fire. In order to maintain flames near these lower limits of viability, one needs to make use of a fuel that is quite volatile. Some form of gas (like methane) or flammable liquid (like paraffin or methylated spirit) is required. If one wants to maintain a fire of leaves, coal, wood, or peat, as would be a more realistic scenario for a primitive culture, the critical size would be far larger. It is a nice coincidence that coal, wood, or peat fires have to be of a minimum size in order to maintain the ignition temperature of the ingredients under typical atmospheric conditions; and that minimum size is just about what is required to keep a human being warm in a natural shelter of convenient size.

These considerations place a restriction upon how small one can be and still make use of fire. If there were no limit on how small a flame could burn in air, then very small creatures could use fire to supply warmth, initiate technologies, and change their environments. But because there is a smallest flame, very small creatures are faced with approaching unmanageably large fires if they are to sustain them with fuel. Their inability to control fire is not only crucial in preventing them from developing various forms of technology; it also restricts their diversity. They cannot spread into regions where climatic fluctuations are large; they cannot populate regions where the mean temperature is very low; and their activities are restricted to the hours of daylight if their light-gathering sensors respond only to visible light.

The use of fire by humans is universal. One finds evidence for the systematic use of fire a hundred thousand years ago, and for the exploitation of natural fires nearly one and a half million years ago. Its principal benefit is the possibility of having a barbecue. Cooking makes food easier to consume and digest, kills harmful bacteria, and enables meat to be preserved for longer. These gastronomic factors serve to enlarge the range of foodstuffs available to fire-making humans, improve their health, and reduce the range over which they need to search for palatable prey. Cooking also stimulates the emergence of a discriminatory sense of taste. Meat can be cooked in a variety of ways; its taste differs from that of uncooked meat.* The nuances of taste that cooking creates and the division of labour that it entails have clearly played a continuing role in human social evolution. Only hominids practised cooking, and, unlike other animals, we take trouble to make food look nice as well as taste good. Thus the ability to make and control fire opens up other evolutionary pathways to a species. It alters the range of foods it can exploit, and the nutritional benefits to be gained from them. It increases the length of the waking day, provides security against predators, and provides a means for clearing land and stampeding animals. There have also been suggestions that the transition to a diet of high-quality, easily

---

* We still identify cooking and eating habits as a form of social distinction. This is not a recent innovation: the Algonquin Indians of north-east America disparaged their northerly neighbours by calling them 'raw meat eaters', that is 'Eskimos'.

digestible food that cooking precipitated may have played a role in the rapid evolution of the brain. The human body displays two anomalies with respect to its overall size: the human brain is relatively large while the human gut is very small. The latter is a sign that the human diet did not require vast digestive capabilities. Thus energy was available to underwrite the expensive expansion of the brain. This might not have been possible if lots of energy had been required for digestion.

One of the most important developments in human history was the innovation of the written word, together with the use of papyrus, paper, and other lightweight materials for its representation and storage. To make use of materials like paper, it is necessary to be large. Small creatures like flies and lizards exploit the adhesive forces between molecules because these surface forces are stronger than gravity over very small areas. Larger creatures cannot use adhesive surface forces to beat gravity because their weights are too great. But, those same surface forces that help tiny creatures to defy gravity ensure that they cannot manipulate surfaces. You cannot turn the pages of a book, however small it may be, if you stick to the surfaces of its pages. Of course, one can also conceive of strategies which one could implement to overcome this problem *today* (cover the pages with some special transparent detergent that reduces the surface adhesion dramatically, just as washing-up liquid stops fat sticking to plates). But such a complex situation would not develop spontaneously, or exist as the first step in the development of information-recording media.

The existence of surface forces is the reason why there is a sharp division between the way living things behave on either side of a dimension of a few millimetres. Above this divide, gravity holds sway, keeps our feet firmly on the ground, and, ultimately, places a limit on how big we can get. Below that scale, life is dominated by the presence of adhesive forces that stick surfaces together and overwhelm the force of gravity (Figure 3.12). In the vicinity of the dividing-line, the balances that are possible between gravity and adhesive forces give rise to a vast profusion of living things. This region offers a striking diversity of possibilities for survival, and unpressured evolutionary niches. Whether you choose to walk on water, or on the ceiling, or ride around on the skins of other animals, all these lifestyles are possible only where intermolecular stickiness can match the strength of gravity. This submillimetre world is not without its drawbacks, though. Wheels are less than useful: their surfaces would feel the pull of surface forces, and they would roll as if the brakes were on all the time. They would not be adaptive.*

---

* None the less, although the wheel seems integral to our own culture, there have been advanced cultures, like the Maya (AD 750), who never invented it. I know of no example where the wheel has evolved in living organisms. The closest approximation is the molecular propeller provided by the flagella of bacteria.

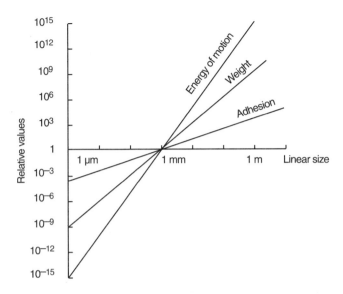

3.12 The variation in energy of motion, the strength of surface adhesion, and weight with mean size. The scale where the three strengths coincide is densely populated by tiny organisms.

Far from the madding crowd: the size of populations

The meek shall inherit the Earth.

St Matthew

Size is a key to survivability. Small animals are common; large ones, especially ferocious predators, are rare. And, if we look at particular ecosystems, we find that animal sizes do not vary continuously over all possibilities. They seem to cluster around definite rungs on an increasing ladder of sizes. This stratification reflects the predatory nature of animal existence: an equilibrium has been reached in which, generally speaking, every creature fits into the mouth of a larger one, and feeds off others small enough to fit into its own. The same pattern of increasing abundance with decreasing size is found throughout the living world until organisms become so small that complete structural redesign would be necessary for them to evolve in the direction of even smaller sizes (Figure 3.13).

At first one might think that this downward trend in the abundances of larger creatures is entirely geometrical. There must be more small creatures than large ones, simply because you can make more small things than large ones from the same amount of living tissue. But this is not a sufficient explanation. If we look

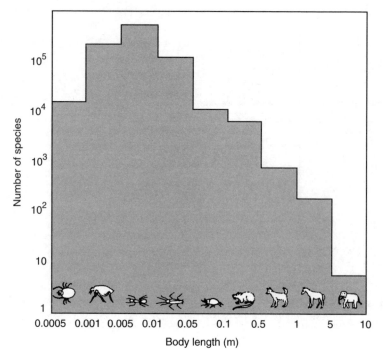

3.13 A census of terrestrial species versus size.

at how the total biomass is invested by Nature across the size-spectrum of living things, we discover that the 'small is best' tendency becomes even more impressive (Figure 3.14). The investment strategy of natural selection is to put its resources in the bodies of plants, and in small animals rather than large ones. The Almighty does seem, in J. B. S. Haldane's words, 'to have had an inordinate fondness for beetles'.

We have already seen that the intrinsic strengths of the forces of Nature determine the maximum sizes to which living things can grow on the surface of a life-supporting planet. But why is the planet not full of large creatures exploiting that upper size-limit to the full? What determines how close they can get to the limit, and in what abundance they are likely to do so?

One constraint is supplied by the ubiquitous second law of thermodynamics. This, the reader may recall, is the scientific underwriting of the familiar experience that things tend to go from bad to worse. It states that, in a closed environment, disorder can never decrease. The reason for this one-way street is simply that there are so many more ways in which a system can evolve from order into disorder, rather than vice versa, that it is overwhelmingly probable that whole systems will tend to become increasingly disordered.

Energy can be neither created nor destroyed; but it is inevitably degraded into

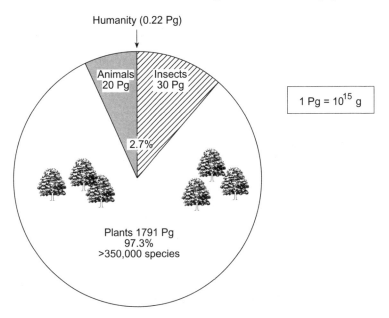

3.14 The composition of the biosphere. The total mass of living material is estimated to be 1841 petagrams (where 1 petagram = 10^{15} grams is roughly equal to the mass of a cubic kilometre of water). Of that total, only 4 Pg is found in the oceans despite the fact that they cover two-thirds of the Earth's surface area, the rest is found on land. It is divided between animals, insects, and plants in the amounts shown. There are estimated to be more than 1.2 million species of living creatures (excluding at least 100 000 micro-organisms), of which at least 800 000 are insects. As can be seen, the biomass is totally dominated by trees (97.3 per cent) and the mass of humanity is negligible (0.01 per cent) compared with that of other animals and insects.

less and less useful forms. If one sets up an industrial process in which the output from one stage is used to power the next, then the ratio of usable energy output to energy input will decrease with each successive stage. Perpetual motion machines are impossible. In practice, we can break the cycle of degrad-ation by injecting some highly ordered energy (like electrical power) into the process at some stage, but that means that the system under consideration is no longer closed. These thermodynamic strictures apply with equal force to the energetics of the living world. We can regard the biosphere as a production line in which a vast abundance of plants is eaten by insects, who are in turn con-sumed by larger ones, who are themselves preyed upon by small animals, who provide lunch for larger ones, and so on. At each level of this pyramid, the available food energy is divided between wastage, maintaining living processes, and producing offspring. Only a fraction of the energy entering a level of the food chain remains for the predators who feed upon it. Each level of the food chain acts like an avaricious middleman: taking its rake-off from the energy resources that it receives before passing them on. As one moves up the pyramid, and enters the domain of the larger predators, there are not so many calories left. The second law of thermodynamics is degrading the energy in the food chain at each link. Thus one sees that the larger creatures are faced with the end of an ever-thinning wedge (Figure 3.15).

Big animals at the top of the food chain are using only a small fraction of the food energy beneath them in the chain, and so they cannot be as abundant as their smaller prey. The relative abundances of animals of different sizes reflects the small fraction of the food energy that they have access to in the link of the chain below them. Moreover, we see that larger species are caught between a rock and a hard place because, as species become bigger, they are faced with preying upon increasingly ferocious, or nimble, creatures just smaller than themselves. To catch them, they require an increased investment of scarce resources in offensive weapons, and they must indulge in more energetic behaviour. A cheetah may be fast, but it lives very close to energetic bankruptcy because its high-speed chases are so often unsuccessful, and consume profligate amounts of energy.

Thus, we see why there will be an upper limit to the sizes of predators. Dragons face a law of diminishing returns. The food resources available to them eventually diminish below that required to sustain them. For this reason, the abundance of different species decreases as their sizes increase: their abundance is determined by the number of food calories available to them at their level of the food chain. Exceptions to this general argument are few, and reflect the adoption of an unusual strategy to bypass levels of the food chain. Elephants

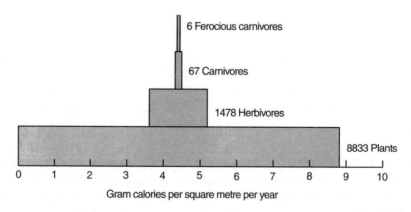

3.15 The food-chain pyramid of a particular environment. At the base are species of plant life deriving energy from photosynthesis. Their calorific value is exploited by herbivorous creatures, who are in turn preyed upon by carnivores of increasing size and ferocity. Data gathered by H. T. Odum in Silver Springs, Florida.

and giant pandas feed on plants, and so cut out the middlemen in the form of small intermediate animals. Even so, pandas spend most of their waking hours eating simply to survive. Their source of food is bamboo, which is unique within their habitat because it is available all the year round. It is interesting that pandas have the teeth of carnivores, and may once have been meat-eaters or omnivores who only survived by adopting a new plant-eating strategy at the base of the food chain. Large baleen whales also feed at the bottom of the food chain, but by means that are not available to land-going creatures. By sieving vast volumes of water, they can extract large quantities of krill and shrimp without exerting huge amounts of energy in hunting; nor do they use energy escaping from natural predators. Their only enemy is man. Their food supply also displays a huge abundance, and replenishes itself very rapidly. Some authors, such as Paul Colinvaux and Beverley Halstead, have claimed that dinosaurs of legendary ferocity, like *Tyrannosaurus rex*, actually lived relatively inactive lives in order to conserve precious food calories. They avoided expending their energy in chasing agile prey by focusing their attention upon disabled animals and carrion. Eventually, they were to lose out to faster, smaller creatures, who were far more efficient in cleaning up these easy pickings. This argument seems weak. Dinosaurs like *Tyrannosaurus rex* do not have the biomechanical design of lumbering sloths; rather, they seem equipped to run at speeds up to 65 kilometres per hour and walk at up to 16 kilometres per hour. Nor do their enormous teeth and jaws seem like the end-point of an adaptation for a scavenging existence. They look like carnivores, and they had plenty of time to lose these accoutrements of the carnivorous life-style during their long period of successful adaptation to the environment, if they had ceased to aid survival and fecundity—an adaptation that seems to have failed only when faced with overwhelming environmental change which eradicated the majority of living things. There are other possibilities that might serve to explain the puzzling fact that, when the age of the giant dinosaurs ended, they were never succeeded by equally large carnivorous mammals. Perhaps being bigger and fiercer just became thermodynamically impossible in the new situation in which the spectrum of smaller creatures had changed.

Regardless of the dinosaurs, our general thermodynamic argument shows why calories become increasingly scarce as one moves up the food chain. Eventually, the food calories available will fall below subsistence level for the way of life needed to gather them. The size of the largest carnivores will therefore depend upon the percentage that each predator extracts from the food chain, and upon the total amount available at the base of the chain. The extraction efficiencies do not change very much as one moves up the chain, and are ultimately determined by invariant aspects of biochemistry. The overriding factor is the amount of usable energy available in the plants at the base of the pyramid. This sets the maximum to which usable energy reserves can stretch.

The foundation stone of the entire pyramid of life is the amount of solar energy available on the Earth's surface, coupled with the efficiency with which it can be incorporated into plants by the process of photosynthesis. On average this process is very inefficient. Only about one per cent of the incoming solar energy is used to produce sugars in plants.

The reasons for this gross inefficiency—twenty or thirty times lower than that of good man-made machines—are various. Only a fraction of the Sun's rays falls in wavebands that are energetic enough to initiate photochemical reactions. The rest do nothing more than slightly warm the surfaces of plants. The intensity levels in the various wavebands received by terrestrial plants are determined by the internal astrophysics of the Sun and by its distance from the Earth. But the weak link in the entire photosynthesis chain, which is responsible for the inefficient use of solar energy by plants, is the lack of the raw material that photosynthesis uses to make food sugars: carbon dioxide gas. Only 0.03 per cent of our terrestrial atmosphere is in the form of carbon dioxide. This is the bottleneck that prevents more solar energy entering the food chain. Even if the intensity of sunlight were greatly increased, the efficiency of sugar production would barely change, because there is not enough carbon dioxide to exploit the extra sunlight.*

Thus, because of the scarcity of carbon dioxide, the total food energy available at the base of the food chain for predators to take a slice at the top of each level is just one per cent of the total solar energy falling on the Earth's surface. Ultimately, the maximum size of animal predators, and their sparsity, is a reflection of the dearth of carbon dioxide in the Earth's atmosphere.

These considerations reveal more than the reason why large animals are rarer than small ones. The need for animals to extract food from their environment, by preying upon smaller ones, ensures that large animals also need to hunt and forage over a wider range. As a result, the population density of animals would be expected to decrease with their size. And indeed it does, as can be seen from Figure 3.16.

If we look at the recent distribution of large carnivorous animals, we find that they ranged over whole continents (and sometimes over more than one continent) before human intervention became a serious hazard for them (see Table 3.1).

This demographic trend creates another problem for large animals: they need to be widely dispersed in order for there to be enough prey to satisfy the energy needs of each; but, if members of a species are too thinly spread, they will not meet potential mates frequently enough for a viable population level to be maintained. Since large animals tend to have small litters and devote long periods of time nurturing their young to the age of fertility, they are doubly prone to the

---

* As one might expect, when plants are grown under artificial conditions, with more carbon dioxide added to the air, they make more efficient use of the incident sunlight.

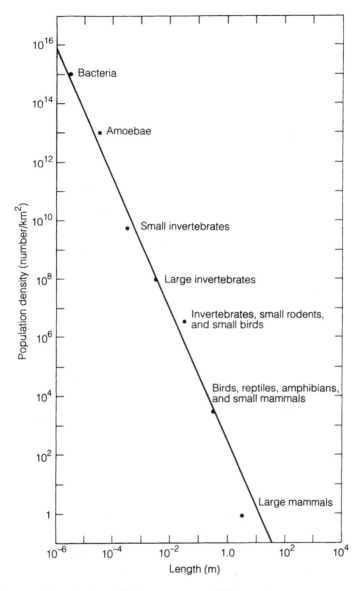

3.16 The population density of living creatures of different sizes.

pressures of low population densities. On islands, or continental land-masses, where limits on the available hunting grounds may be placed by lakes or mountain ranges, the rarity of large predatory animals is likely to be exacerbated by the conflicting constraints imposed by the need for adequate breeding opportunities and sufficient food supplies. They combine to make the survival of large animals rather precarious.

Table 3.1 The demographic ranges of ferocious predators

Lion (*Panthera leo*)	Balkans and Arabia to central India, nearly all of Africa
Tiger (*P. tigris*)	Much of Eurasia
Leopard (*P. pardus*)	Much of Africa and Eurasia
Jaguar (*P. onca*)	Southern United States to northern Argentina
Snow leopard (*P. uncia*)	Mountainous areas from Afghanistan to Lake Baikal and eastern Tibet
Cheetah (*Acinonyx jubatus*)	Middle East to central India, Africa except for the central Sahara and rainforests
Cougar (*Felis concolor*)	Most of North America to southern Chile and Patagonia
Spotted hyena (*Crocuta crocuta*)	Sub-Saharan Africa except in rainforests
Coyote (*C. latrans*)	Most of North America
Gray wolf (*Canis lupus*)	Most of Eurasia and North America
Hunting dog (*Lycaon pictus*)	Most of Africa
Asiatic black bear (*Ursus thibetanus*)	Most of central and eastern Asia
American black bear (*U. americanus*)	Most of North America
Brown bear (*U. arctos*)	Most of Eurasia (except tropical regions), northern Africa, most of North America
Polar bear (*U. maritimus*)	Arctic Eurasia and North America

Increasing size also leads to inflexibility and over-specialization. Although large size successfully insulates organisms from small changes in their environment, it puts them at risk from major ones. When disaster strikes, they take the longest to recover because of their small litters, and the fact that the sexual reproduction time increases with the size of the animal (Figure 3.17).

Their lengthy reproduction cycle means that large creatures change more slowly than small ones, because genetic changes can occur only during the single-celled stage of the life cycle. Many more small changes are necessary to produce an appreciable effect upon a large animal. As we go from the Equator to the Poles, we see that animal diversity decreases with predictability of the climate. Seasonal changes become more severe and abrupt; rapid freezing and thawing of water becomes common and erratic, just as it does as one moves up a mountainside. In the foothills, life is still relatively diverse but, as one ascends, the increasing severity and unpredictability of temperature changes leads to less and less variety. In general, changeable or dangerous environments favour organisms that produce many offspring and have short generation times. By contrast, benign environments favour organisms with few offspring and long generation times, whose young can be placed in favourable ecological niches that they are well equipped to exploit. The relative vulnerability of large animals to the vagaries of a rapidly changing environment means that smaller creatures tend to be most likely to survive climatic revolutions. Consequently, they dictate the underlying rate of evolutionary change. The pattern for large predators is shown in Figure 3.18.

Let us return to the puzzle of why the large dinosaurs were not followed by

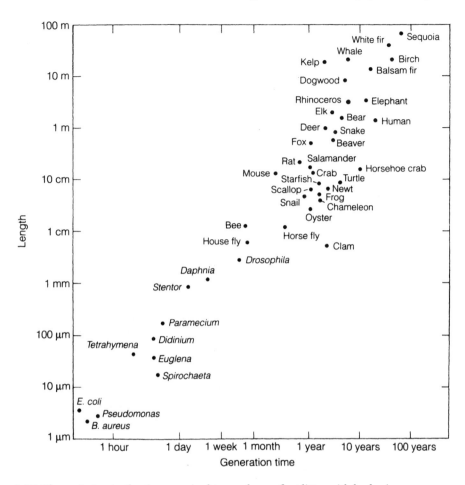

3.17 The variation in the time required to produce a first litter with body size.

equally large carnivorous mammals. We might appeal to the conflicting pressures of sparse food supplies, or the need to maintain population densities at a high enough level for breeding, as a limit to the evolution of such large meat-eating mammals; and we might seek some peculiarity of the dinosaurs that enabled them to evade the full force of these limits. Perhaps they had much faster population turnovers or more efficient digestive systems than large mammals? Perhaps young dinosaurs could eat a wide variety of small animals and large insects, so widening their access to the lower reaches of the food chain? This would distinguish them from present-day carnivores, whose young eat the same diet as their parents. Another possibility is that the dinosaurs had many more young than would be expected by extrapolating the trend from what we know of large mammals. We know that the number of young produced in each litter by

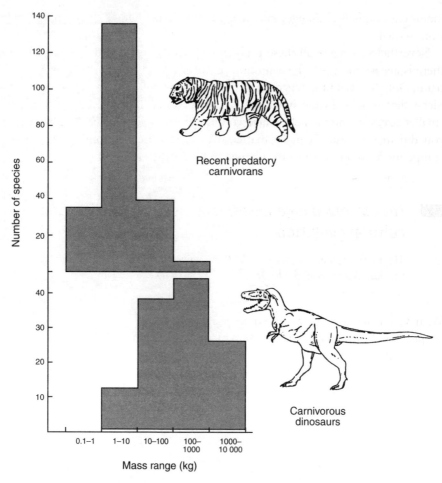

3.18 The number of extant and extinct large predators versus body weight.

contemporary land animals decreases with body size, but large birds that nest on the ground do not follow this trend. Their clutch sizes do not vary significantly with body size. As a result, birds have far greater potential for reproductive success than do similarly sized mammals. Could meat-eating dinosaurs have followed the same trend? Alternatively, it has been suggested that their metabolism was more efficient than that of mammals, allowing them to make better use of their food supplies. At some fossil sites the range of skeletal remains found in dinosaur habitats suggests that their food needs may have been considerably less than those of large mammals. While any one of these factors might be sufficient to explain the preponderance of large dinosaurs compared with mammals, it is also possible that all of them combined in complicated ways to tip the scales in ways that permitted the dinosaurs to continue their precarious existence until major environmental changes intervened to eliminate them. The question is far from settled.

Nevertheless, despite all these problems of being large, the only place where there is always room for the evolution of novelty is at the top of the size spectrum. Only by becoming bigger than the biggest extant animals can one enter a niche that is not already inhabited by competitors. If you evolve towards a smaller size, you enter a niche where you must prey upon smaller creatures than you did in the past. To make matters worse, you are confronted by intense competition from those already adapted to that niche.

Les liaisons dangereuses: complexity, mobility, and cultural evolution

> The mind of man is capable of anything—because everything is in it, all the past as well as all the future.
>
> Joseph Conrad

We have learned that, as we scan the size spectrum of living things, the number of species to be found decreases with increasing size. But this decay in diversity is compensated by a growth in the complexity of these species as their size increases (Figure 3.19). Passing from small organisms to large ones, we find a steady increase in the number of different types of cell that are present in their

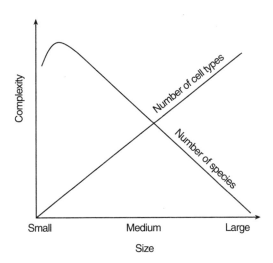

3.19 The pattern of variation in complexity with size of organisms. Internal complexity is gauged by the number of cell types present in the organism, and external complexity by the number of different species.

bodies—a reflection of the gradual subdivision of function that is associated with the evolution of organized complexity.

Since these different cell types are of roughly equal size, the overall size of an organism is controlled by the total number of cells. By banding together in large numbers, cells avoid competition from other small organisms. They explore a new niche, which often offers the best strategy for transmitting their genetic information to the future. The passage from small living things to large ones reveals a gradual transfer of diversity from the realm of external appearances—the range of different species that exist—to the internal make-up of fewer species. This correlation between size and complexity suggests that evolutionary selection for one of them led to increase in the other.

It is not known whether there is any limit to the number of cells that could make up a functioning organism. There probably is, if only because of the constraints imposed by the need to maintain some thermal equilibrium and connectivity between parts of the organism. To see why, consider the problem of building an artificial 'brain', and ask if there are limits to its size and capabilities. At first, you might think that the bigger the brain the better. But, stop and think what computers do, and what *you* do when you think about what *they* do. Each computational step processes information, does work, and produces waste heat—just as the second law of thermodynamics demands. If we build a larger and larger artificial brain, then the volume of its circuitry will grow faster than the area of the surface enclosing it from which the waste heat can be radiated away. We are back with our old dilemma: the competition between volume and area. If everything is simply scaled up in size, the 'brain' will eventually overheat and melt. To overcome this problem, we could take a leaf out of the book of Nature and give the computer a crenellated fractal surface structure, so as to boost its area relative to its volume.*

There is, however, a price to pay for this strategy, too. To keep all the parts of the computer near an irregular surface interconnected will require a far greater length of circuitry. This means that the computer will operate more slowly. More time will be needed to coordinate signals sent from one part of its surface to another. There seems to be a trade-off between increasing volume, computing power, surface cooling, and processing speed. Perhaps there is some ultimate limit on how big, or how powerful, a computer can be? So far we do not know.

Likewise, if we look to small sizes, there is evidently a minimum number of cells for a living thing to function or respond to the pressures of natural

---

* The worst case is provided by enclosing the volume of the 'brain' within a sphere, for this gives the smallest possible surface area that could smoothly enclose that volume. Remarkably, there is in principle no limit to how large the enclosing area can be if its surface is sufficiently irregular. In practice, the smallest scale of the superficial irregularities would be limited by the sizes of atoms (10^{-8} cm), or their nuclei (10^{-13} cm).

selection. In our thought-experiment to construct an artificial brain, we would find that the pulsing frequency of the central processor would need to increase if the computer shrank in size, in order to maintain the required power output. Now, the surface would need to be as spherical as possible, or very well insulated, in order to minimize power losses to the outside. Eventually, the stresses imposed by the processor (or more likely the intervention of physical effects associated with short-range interactions by other forces of Nature) would intervene. This same effect limits the smallness of animals and birds. Their pulse rates increase with falling size, and any bird significantly smaller than the smallest humming birds would become anatomically impossible because of the enormously high pulse rate that would be needed to maintain its body temperature.

▓▓ Network news: branching out

> All great truths are obvious truths. But not all obvious truths are great truths.
>
> Aldous Huxley

In recent years there has been renewed interest in the ways in which many attributes of living organisms vary with their mass, M. The principal puzzle was the appearance of one-quarter powers in the scaling relationships. We find, for example, that the rates of cellular metabolism are proportional to $M^{1/4}$, life-span to $M^{1/4}$, and the whole organism's metabolic rate to $M^{3/4}$. There are well over a hundred observed scaling 'laws' that have multiples of a ¼ power in them. The puzzle is that if you were trying to predict these relationships ahead of seeing the data then you would have expected the ubiquitous power to be ⅓ rather than ¼ because these relationships are expected to scale with size, which is proportional to $M^{1/3}$, as mass equals the product of density and volume and the density of living matter is constant. It is as if nature was really operating in a four-dimensional world rather than a three-dimensional one.

Geoffrey West, James Brown, and Brian Enquist first proposed that all these scaling laws could be explained if we assumed that living organisms were fractal networks that distributed resources from their largest scale down to their smallest (dictated by capillary size, which is the same in all organisms irrespective of size) in an optimal way so as to maximize the area across which they can absorb and release nutrients and minimize the time needed to transport them around the organism. The fractal network produces the efficiency that we would get by adding an extra dimension to space. If we draw a straight line on a sheet of paper then it just covers a one-dimensional path. But draw a scribbled wiggly line all over the page and the one-dimensional line can almost cover a whole

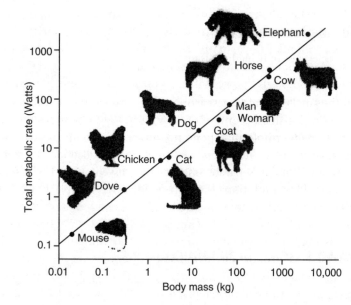

3.20 Metabolic rate increase with the three-quarters power of body mass for a wide range of animal sizes.

two-dimensional area. The wiggly line behaves as if it was a surface. This is the effect of the fractal network in living things. This simple consideration explains the ubiquity of the ¼ powers in all the biological scaling laws.

Despite the success of these simple considerations there have been others who believe that fractal networks are not a necessary part of the argument. Jayanth Banavar, Amos Maritan, and Andrea Rinaldo argued that the ¼-power rules follow from a simpler model in which resources of any sort flow outwards from some source to a variety of take-up points. They argue that any network for distributing nutrients has a circulation length L which serves about L^3 sites where nutrients are used. Any nutrient would pass L of these sites on the way to this end-usage point. So the total of nutrients in the system at any given time must be roughly given by the number of final destinations times the number of stops on the way, which is proportional to L^4 and not L^3.

At the moment these two simple arguments give the same welcome conclusion, but they are based upon different models of the organism. The resource flow model doesn't need to assume fractal structure, but it assumes implicitly that the organism's internal network fills up its volume so efficiently that it has an effectively four-dimensional internal volume, just like the fractals of West *et al.* There is probably one final step yet to be taken in this debate to clarify the relationship between these two arguments or to subsume them into something that is slightly more general.

The Go-Betweenies: messing with Mister In-Between

'The Earnestness of Being Important'

The Little Book of Stress

In between the smallest and largest living things, we enter a world of growing complexity. As organisms become more complex, they rely upon increasingly delicate forms of intermolecular bonding, and more complicated molecular shapes. The investment of resources in a single large collection of interlinked cells, rather than in many separate small organisms comprised of a few cells, would be a short-lived evolutionary experiment if large complexes were invariably at greater risk to infelicitous fluctuations than small ones. Fortunately, the opposite is true. One advantage of being composed of a very large number of components is that random fluctuations in their distribution and functioning decrease inversely in proportion to the square root of the number of components. If the system is too small, then it will suffer from relatively large random fluctuations, and will probably fall victim to a fatal lack of fidelity in its genetic-copying programs. Only if it passes this restriction will it have the opportunity to evolve higher forms of complexity that include (like our bodies, and some computer programs) systems to correct genetic copying errors. Of course, it pays to invest scarce resources only in particular types of repair and error correction. Whereas the resources used to heal small cuts and grazes are well spent—offsetting the risk of fatal infection at an early age, before offspring have been produced—regenerating replacements for amputated limbs is not.

If we examine how the brain size of different species of living creatures (not different members of the same species) varies with their body sizes, we find that there is a direct trend, as shown in Figure 3.21.

The exact slope of the graph is the subject of much discussion, which has yet to produce a persuasive explanation. It shows that, roughly,

$$(\text{Brain weight}) \propto (\text{Body weight})^{3/4}$$

However, the folded structure of the brain, together with the importance of its links to the nervous system, may mean that its weight (or its volume) is an inappropriate indicator. Its surface area may be more important. If intelligence does increase with the evolution of brain size, then both seem to be unique evolutionary traits. Although there have been larger, faster, and stronger animals in the past, none have been as intelligent as some of those living today. A close examination of how brain size varies with body size displays a fairly continuous spectrum of increase for land-based creatures up to a certain large size; then there is a gap, with only *Homo sapiens* beyond it. The intervening examples seem to be missing. Perhaps they were eliminated early on by the aggressive tendencies

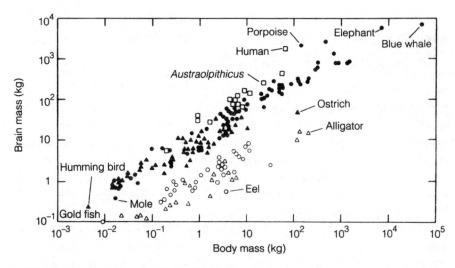

3.21 The variation in brain weight with body weight.

of *Homo sapiens*. We know that there were once other intelligent species, like Neanderthal man and Cro-Magnon man, but the injuries found to many of their fossil skulls suggest that they may have been removed through conflict with *Homo sapiens*. By contrast, if we look at the spectrum of marine life these gaps do not arise. Life in the sea appears far less competitive. There is less pressure upon food resources and on territory. But if there is so little pressure in the marine environment, why does the population of the sea not dramatically increase?

Perhaps becoming big is the only way in which it is possible for the brain to become big, and hence for intelligence to increase. Some biologists, like Stephen Jay Gould, have argued that such a specific ability as language could be merely a by-product of enlarged brain size. But this seems a strange argument. Increased brain size is a risky and costly evolutionary pathway. It would prove cost-effective only if it offered some dramatic advantages. Language is the most impressive of these advantages. It is much more likely that large brain size evolved as a by-product of natural selection for enhanced linguistic ability (perhaps by sexual selection, because loquacious individuals were more interesting and hence appealing, as Geoff Miller has suggested) than vice versa.

If we merely doubled the size of a brain, without changing the nature of its neural connections, then the increased size would not signal a doubling of capabilities. Large animals will tend to have larger brains than small animals because of the overall scale of their body engineering. To allow for this, we re-examine the relation between brain size and body size, but remove the increase in brain size that arises solely from the growth in body size. What remains is called the encephalization quotient, EQ (defined to be the brain size of the mammal concerned divided by the average brain size of all mammals with the same body size): it shows whether brain size outstrips the level one would expect for a given body size. As you might expect, human beings are vastly over-endowed with grey matter when viewed in this way. Our nearest rivals, dolphins and porpoises, are rather puzzling. In a general way, we can understand how complexity and intelligence—and hence EQ—needs to increase with size. Our discussion of the thermodynamics of the food chain showed how difficult life becomes for large predators. Not only do calories become scarcer, but the prey on which they feed become correspondingly rarer, fiercer, and more agile as they get larger. More and more of a large predator's resources need to be devoted to programming a sophisticated guidance system that enables it to hunt mobile prey efficiently. You are what you eat. In the case of dolphins, it is hard to see what their large brains are for. They have abundant sources of food, which is fairly easily caught, and they do not appear to be unduly threatened by preda-tors, because of the sparsity of large sea creatures. Perhaps it has something to do with their sonic guidance system? Recently, dolphins have been seen to display aggressive behaviour towards smaller porpoises who may have been creating too much noise for the dolphins' sonar to work efficiently.

If we return to the puzzle of our own enhanced brain size then, clearly, an increase in body size is not enough to explain it. Nor is it really necessary. The introduction of a few advantageous genes that prolong the portion of youth during which the brain grows can give a species an anomalously large brain in relation to its body size. And, indeed, this appears to be the case if human growth is compared with, say, that of our nearest genetic relatives, the chimpanzees.

The pay-off from the evolution of human brain-complexity has been the possibility of development, adaptation, and the avoidance of competition by non-genetic means. By passing on ideas through social interaction, by means of language, records, images, symbols, gestures, and sounds, our development has proceeded far more rapidly than by encoding particular types of information in genes. The information that can be passed on by these behavioural and cultural means is of a type that cannot be transferred by genetic inheritance. It enables learning, teaching, and knowledge to accumulate. Whereas information transfer by genetic means is limited to inheritance by the offspring of an individual, the influence of ideas and culture is limitless in its potential range. Favourable adaptations can be spread through the population very rapidly. An idea, like taking physical exercise or avoiding unhealthy food products, which can be seen to enhance the chance of survival can become a common possession almost over-night, once propagated by mass media. Cultural transmission allows detailed information about the local environment to be passed on quickly, and is essen-tial for survival in an environment that changes more rapidly than the time interval between successive generations.

In the previous chapter, we saw something of the explanatory power of natural selection. Wherever we find intricate interwoven complexities, we find the hand of time, slowly fashioning adaptations. In this chapter, we have seen how factors common to all organisms produce constraints upon their development that limit the variations upon which natural selection is free to act. Size is a pervasive influence that determines aspects of an organism's structure and strategy for survival; it is linked to habitat and habit, to lifestyle and quality of life.

Despite the impressive achievements of the past four thousand years of human history, that period is a relatively brief interlude in the span of human existence. The huge number of generations that humans spent hunting and gathering might, therefore, hold clues to the origin of our instinctive behaviours, likes, and dislikes. By examining the early environment in which humans lived and evolved for such vast periods of time, we should find clues to the selection for adaptive behaviour and the elimination of maladaptive behaviour, which have left an imprint that we still bear today. Although we have cautioned against believing all adaptations to be perfect, or all traits to be optimal, some human mental abilities may be the results of adaptation to primitive circumstances. The enormous periods of time during which our ancestors were foragers and hunter-gatherers in the Pleistocene epoch, between two million and ten thousand years ago, are likely to have been formative for our species (see Figure 3.22).

While it has long been fashionable to regard each human mind as a blank slate that is informed by learning only after its birth, this view has been found woefully inadequate to explain the development of human language (see pp. 26–30). Genetic pre-programming by natural selection endows us with human linguistic abilities. Those in possession of such abilities, even in a more primitive form, had a clear survival advantage over those that did not: linguistic ability is adaptive. Whenever we find widely shared human features and behaviours—especially those of great complexity—it behoves us to look for a possible adaptive explanation. In practice, it is the *differences* in behaviour between one person, or group, and another that seem most easily attributable to learning.

Our Pleistocene hominid ancestors lived mainly in tropical climates with quite distinct environmental conditions. The hours of daylight varied little throughout the year, and roughly equalled those of darkness. There was very little seasonal variation in temperature. What little there was would be negligible compared with the diurnal temperature variations; these could be considerable, because an absence of cloud cover allows very rapid radiative cooling. These tropical latitudes also had only modest winds and negligible wind-chill factors, but gave high exposure to ultraviolet radiation, and a risk of dehydration. The high daytime temperatures ensured that the key to survival was the rainfall—

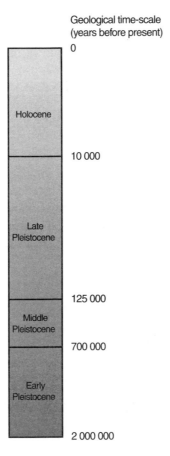

Geological time-scale
(years before present)

0

Holocene

10 000

Late
Pleistocene

125 000

Middle
Pleistocene

700 000

Early
Pleistocene

2 000 000

3.22 The time-scale for the Quaternary, the past two million years of geological time.

and most critically, the *lowest* rainfall levels during the dry seasons. Rainfall is highly variable in tropical climates, with long dry seasons followed by torrential deluges. The length of these dry seasons is the most influential factor in determining the diversity of vegetation. When considered in combination with the mean annual rainfall, it helps us to create a picture of how different habitats emerge (see Figure 3.23). As the dry seasons lengthen, the vegetation becomes simpler and poorer. These variations also influence the spectrum of living things that live off them, the time they must spend foraging for food, and hence the patterns of behaviour they adopt.

Our own physiology displays remnants of an early adaptation to tropical environments. We have a greater sweating rate and far less body hair than other mammals. These features helped to regulate our body temperatures very efficiently, even when participating in hunting activities, which produce

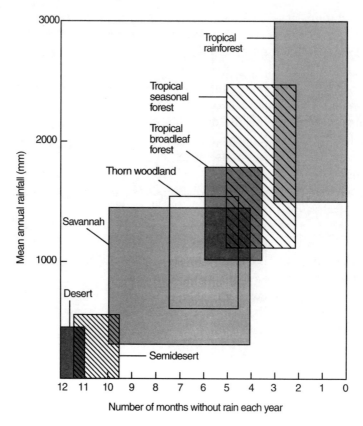

3.23 The characterization of various habitats by the mean annual rainfall and the length of the dry season.

considerable heat stress and water loss. Here, our size begins to play a role. Small animals need to eat a far larger fraction of their body weight to meet their daily energy needs. They must choose foods with very high calorific values, or else spend most of their lives feeding. Therefore, they must find reliable, long-lived food supplies, hibernate, or invent ways of storing food for hard times. Large animals are not so limited. They are more mobile, and can sustain themselves on patchier resources. Bipedalism aids mobility, and endows humans with remarkable long-distance endurance and flexibility of movement in widely differing terrains. If there is evolutionary pressure to discard incessant foraging for unpredictable resources, then there will be adaptive advantage in large size and mobility. There are penalties to evolution in this direction, which must be outweighed by these advantages. Increased size and weight makes the land a safer, easier environment than the trees.

The environment in which the development of larger land-going hominids seems to have occurred was that of dry open-savannah grasslands, with only sporadic tree cover. As we saw in Figure 3.23, this is an environment with limited, but highly variable, rainfall. Survival requires adaptation to the problems posed by such an environment. The poverty of the vegetation in the drier season requires much more diverse searches for food, and probably led to the introduction of meat into the diet. Hunting is a challenging activity, which selects for increased cooperative abilities and higher intelligence. It also encourages social interaction. As well as the need for groups to hunt large and dangerous prey, there is also the possibility of sharing the large quantities of unstorable food that each kill provides. Grains and berries can be kept; meat cannot. Meat-eating is a far less specialized source of sustenance than plants and berries. There is an enormous diversity of fruit and plant-life (some of which is inedible or poisonous), but little variation in the forms of meat. Accordingly, herbivorous creatures display a corresponding diversity, which far exceeds that of carnivores. In contrast, the need to be mobile and exploit unpredictable food sources in a varying climate encourages wide-ranging hunting. This makes the hunter adept at utilizing resources from many environmental niches, each of which may be occupied by specialized, but comparatively immobile, local species. Hunters may take relatively little from each niche in comparison with its principal predator, but the total yield will make this eclectic exploitation of resources a very advantageous strategy. Diverse hunting, mobility, and a dispersed population are closely linked. Today, we are very impressed by the intense concentration of people in particular areas of the Earth, and derive great benefits by increasing local population densities. But this was rarely the case in the distant past. When the overall resources are very great and population levels low, it pays to disperse and find a food source that is not being exploited by others, rather than increase demand for limited local resources.

One of the most intriguing correlates with the size of an organism is its lifetime. A large organism, especially one with a large brain, is a considerable evolutionary investment. Becoming large is a strategy that will therefore become extinct if large organisms do not live a long time and use their longevity for some purpose that enhances fecundity or survival.

By its longevity, a large creature could maximize its reproductive output. Yet the number of offspring from large animals is often very low, the gestation period and the period between births is large. Consequently, large creatures lavish far more care upon their young than do smaller, more ephemeral creatures and their mortality rates are lower. Large creatures have sacrificed reproductive efficiency in the interests of greater efficiency in the use of food resources. These strategies require particular social structures. Longevity requires creatures to interact with members of their own species over long periods of time. Lengthy

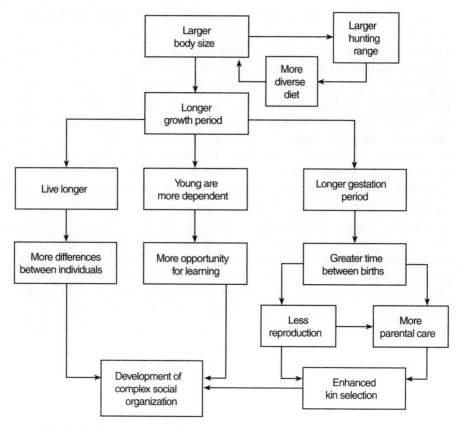

3.24 The chain of consequences that flow from evolutionary increases in body size.

nurturing periods for the young also help to create a complex pattern of social behaviour that makes altruistic behaviour advantageous. These dependencies interweave to produce a network of consequences that flow from increased size (see Figure 3.24). The advantages they present are considerable, and may have played a vital role in the rapid evolution of human capabilities during their early stages.

The rivals: the evolution of cooperation

> One friend in a lifetime is much; two are many; three are hardly possible.
> Friendship needs a certain parallelism of life, a community of thought, a
> rivalry of aim.
>
> Henry Brooks Adams

Natural selection tests and sifts the strategies for social interactions between individuals. It shows that certain 'strategies' of interaction, or of social organization, will lead to greater benefits than others. If adopted, they must be resistant to invasion by individuals and groups adopting variants of them. In this way it is possible for particular, ordered patterns of social behaviour to emerge spontaneously. As we shall see, repeated social interaction with many individuals plays a key role in the spontaneous emergence of stable social structures. We should stress that although the word 'strategy' is employed here, it does not necessarily imply that actions are consciously purposeful (although they could be). It is simply a term to describe a pattern of behaviour employed in the face of competitors, known or unknown. An organism need not 'know', consciously, why one pattern of behaviour is more advantageous than another. A strategy just needs to have better results than the alternatives for it to be selected in the long run because its users will survive more often. One of the most interesting outcomes of such an approach is an understanding of how *reciprocal altruism* can emerge. Cooperative behaviours that offer mutual benefit—I help you repair your car in return for your help in mending my roof—can leave both parties better off, even though they incur costs for actions that do not benefit them directly.

The advantages and pitfalls of cooperation are illustrated by the classic problem of the 'Prisoner's Dilemma'. Two prisoners are held in separate cells and are prevented from communicating with each other. Their gaolers urge them both to confess, telling each of them that if he confesses and his partner does not, then he will go free, while the other will receive the maximum term of five years' imprisonment. If both confess, the confession of each will be worth less to the judiciary, and so they will both receive a three-year prison term. If neither confesses, then both will be convicted of only a minor offence, and will go to gaol for only one year. What strategy should each prisoner adopt?

Consider the position of prisoner A. If his colleague, prisoner B, confesses, then A should also confess, since otherwise he would get five years, rather than three. On the other hand, if B does not confess, then it is in A's interest to confess because then he would go free. Thus, whatever B does, it is in A's interest to confess. Since the same reasoning applies to B, we conclude that the best strategy for each to adopt is confession. But this joint confession results in them both receiving three years, rather than the one year that they would each have received had they kept silent. Nevertheless, it would be contrary to the self-interest of each to keep silent, even though both would be better off if neither confessed. This is the Prisoner's Dilemma; in essence, it is faced by every individual in interactions and social contracts with other individuals; for it is always in the self-interest of an individual to get something for nothing; it always seems to be in an individual's self-interest to cheat another, even though both might be better off if neither cheated. This seems to argue that altruism and cooperation cannot evolve spontaneously in a group of individuals, each of whom pursues his own interests. It would require some dictator—either real or imaginary—to impose patterns of cooperative behaviour upon members of the group. However, this reasoning neglects to include the effects of repeated interaction between individuals in a community. This is one consequence of social complexity which can make cooperation advantageous.

Consider a sequence of interactions (called 'games') of the Prisoner's Dilemma involving two people. The pay-offs from Player A and Player B each adopting policies of cooperation or non-cooperation are specified in the accompanying table describing the four situations.

PLAYER A

		Cooperate	Don't cooperate
PLAYER B	Cooperate	A and B get R	A gets S B gets T
	Don't cooperate	A gets T B gets S	A and B get P

To be specific, choose $R = 3$, $P = 1$, $S = 0$, and $T = 5$. Then, as we have seen, it is in the rational interest of both players not to cooperate, even though this means that they receive a pay-off 1, rather than a pay-off 3, which they would have both received had they cooperated. In general, the Prisoner's Dilemma arises in situations where the pay-offs for pairs of strategies obey the inequalities.

$$T > R > P > S \text{ and } R > (T + S)/2,$$

if both players select their strategy before they know what strategy the other has chosen. But, suppose that this game is played many times over in a large community of players, the total pay-off being summed over all the games. In this case, there is a depreciation factor to be considered. The present value of a potential pay-off in the future is not as great as that of a pay-off now. The game might halt for some reason before the pay-off was received—the player might die, or his bank might collapse. The pay-off from each game is therefore discounted relative to that from the previous game by a discount factor, d, where $0<d<1$. The pay-off expected from a large number of games is obtained by adding the pay-offs expected to accrue from each game, where the pay-off from each game is calculated by multiplying the pay-off of the immediately preceding game by the discount factor, d. For example, if the number of successive games is very large, so that we can treat it as being infinite, then the expected cumulative pay-off to both players if they cooperate in all games approaches closely to the sum*

$$R + Rd + Rd^2 + Rd^3 + .. = R / (1 - d).$$

Thus, cooperation becomes a possible rational strategy that is mutually beneficial, because although a given player does not know the other's choice in the present play, he does know what the other chose in previous plays. He can choose his strategy for the nth play in accord with what choices his opponent has made in the previous $(n - 1)$ games.

The discount parameter, d, measures the importance of the future. Only if d is sufficiently close to 1 (that is, only if the present value of future pay-offs is sufficiently high) is it possible for a 'friendly' strategy, in which the player cooperates until the other player doesn't cooperate, to be a collectively stable strategy; that is, one that, if adopted by everyone, cannot be bettered by anyone adopting a different strategy. In order for a strategy to persist in Nature it must be collectively stable, for there will always arise individuals who will try different strategies. Biologists call collectively stable strategies 'evolutionarily stable strategies'. A population of non-cooperators can, however, be successfully invaded by groups of cooperators if d is large enough and if the relative frequency with which the cooperators interact with each other, rather than with the non-cooperators, is sufficiently high. For example, if we choose $T = 5$, $R = 3$, $P = 1$, $S = 0$, and $d = 0.9$, then a cluster of individuals using a 'friendly' strategy of 'cooperate until the other does not, then don't cooperate for one game, then cooperate until the other does not cooperate again' can successfully invade a population of non-cooperators if just 5 per cent of their interactions are with those not adopting this strategy. By contrast, individual cooperators cannot successfully invade a population of non-cooperators because the strategy of total non-cooperation is also collectively stable. A group of non-cooperators cannot, however, successfully invade a population of cooperators using any collectively stable strategy.

In general, a pattern of response embodying a cooperative strategy can invade a population of non-cooperators if, and only if, it leads to cooperation with other cooperators and excludes (or penalizes) non-cooperators. Any pattern of behaviour violating this rule will end up being selected against. There are many patterns of decision that follow this prescription, and any one of them could, in principle, have been selected for, over huge periods of early human history. One can see that, whereas it is important to be able to detect indiscriminate

---

* This sum is an infinite geometrical progression. To check this, just multiply the pattern of terms on both sides of the equation by $(1 - d)$: the right-hand side is clearly R, while the left-hand side produces a succession of terms that cancel each other out, positive against negative, leaving only R.

non-cooperators (cheats), there is no need to have an ability to detect indis-criminate cooperators (altruists)—because few will survive in the long run. Altruists can always arise because of genetic drift, or be sustained by conscious choice, or the actions of a 'dictator' in the community.

When one comes to examine particular social interactions in ancient times, there is a need to avoid assigning contemporary values of costs and benefits, rather than those appropriate to those in a primitive hunter-gatherer economy. We have inherited a liking for sugary and fatty foods—a fact that the food industry and its advertising agencies remorselessly exploit. Our instinctive liking for such foods is probably a remnant of the huge calorific benefit of these scarce resources for early hunter-gatherers. We now judge their food value differently. It is unlikely that we possess significant adaptive responses to agricultural prod-ucts, since farming has been a human activity for little more than ten thousand years, compared with two million years of hunting and foraging.

This analysis of optimal behaviour patterns might apply to sharing food, exchanging services, exercising care over the young, hunting on behalf of others, and so forth. It also leads us to expect that the tactics of bartering, and of games of strategy, will be activities for which humans possess a transcultural affinity. It shows how social interaction—in particular, the repeated social interaction that derives from longevity and large size in conditions where individualism did not pay—has unexpected consequences for the evolution of patterns of behaviour that appear, at first sight, to run counter to the expectations of natural selection. We should stress, however, that just because certain behavioural patterns are optimal in this way, they are not therefore 'good', or 'desirable'; an ethical pre-scription based upon evolutionarily stable strategies has no special status, and we might well choose to reject it for other reasons.

If we made contact with an extraterrestrial society, we might expect that it evolved from patterns of social behaviour that were once 'naturally' selected in preference to others. We should not be surprised, under appropriate circum-stances, to find cooperative, altruistic behaviour—regardless of the existence of transcendental beliefs in the existence of absolute standards of good and bad behaviour, which we find at the root of most systems of religious belief. Nor, we might add, are such beliefs necessarily undermined by the fact that they are found to coincide with patterns of behaviour that are optimal for an individual's good in some cost–benefit analysis. Alternatively, one could incorporate the pay-offs from correct or erroneous religious beliefs into the overall analysis of pay-offs and strategies. The first to do this, albeit in a limited way, was the French philosopher Blaise Pascal. In his *Pensées*, compiled in the late seventeenth century and published posthumously, he marshalled many arguments in sup-port of Christian religious belief. One displays the first example of a strategic game. He considered someone contemplating laying a bet upon which his future

destiny hinges. Beginning with the agnostic's assumption that 'if there is a God
. . . we are incapable of knowing what He is, or whether He is', and 'reason can
settle nothing here . . . a game is on', Pascal argued that the logical response of
the prudent person to his ignorance is to bet your life on God's existence.* If one
does, then there are two outcomes. If God exists, then belief brings an infinite
reward, and unbelief an infinite loss; whereas, if God does not exist, then
unfounded belief costs at best nothing, and at worst, only a finite loss of time
and effort. On this basis, Pascal's conclusion is that the agnostic should bet upon
God's existence.

It is important to recognize that this 'selfish gene' type of altruism is not quite
what it seems. Although biologists often suggest that this lays a basis for understanding human altruism and the value we place upon it, it falls short in many
respects. It is altruism without an altruistic motive. In fact, it is exactly the
opposite: altruism with a selfish motive. This is the thinking behind reciprocal
altruistic activity of the 'I'll scratch your back if you'll scratch mine' variety. In
practice, we have come to admire altruism which goes far beyond that required
simply to maximize return in a game theory problem. The interesting questions
for anthropologists and ethicists to answer is why we do that, and why individuals do occasionally act in ways that are not in their own interest or even that
of any other individuals who share their genes. Curiously, we recognize that, as
Edward O. Wilson and Michael Ruse admit, 'human beings function better if
they are deceived by their genes into thinking that there is a disinterested objective morality binding upon them, which all should obey'.

■ The secret garden: the art of landscape

> Rock of ages, cleft for me,
> Let me hide myself in Thee.

Augustus Toplady

The fact that our ancestors spent very long periods in tropical savannah habitats
leads us to expect that some of our emotional responses to such an environment
may possess adaptive features. Instinctive aesthetic reactions to the world could
not have evolved if, on average, they contributed negatively to survival. By contrast, those responses that enhance the chances of survival will persist. This is
why rotten meat tastes unpleasant to us, while sugar is sweet. Some of the most
interesting evolved responses are those associated with our responses to the
environment. They provide us with important clues to the source of our most
basic aesthetic preferences.

---

* *Pensées*, No. 223; translated and arranged by H. F. Stewart (Pantheon edn, 1950).

The relative longevity of early humans ensured that they would need various habitats to maintain a life-long supply of resources. Their mobility allowed them to meet that need. Indeed, studies show that early hunter-gatherers moved frequently. The mobility of humans ensures that they will need to make choices about the best environment; the criteria used to make those choices will inevitably be acted upon by natural selection over very long periods of time. Small organisms that are short-lived, or fixed in space, or moved aimlessly by winds and water currents, or limited in their foraging ranges, will not encounter the problem of environmental choice.

As mentioned above, the habitat in which humans originated was that of a tropical African savannah. It is therefore possible that we have developed preferences for environments with many of the characteristic, life-enhancing features that this habitat offered during the Pleistocene epoch. We expect that the propensities engendered by adaptation would dispose us to identify good habitats—both with regard to their present state, and that expected in the future. These will have interesting aesthetic byproducts because our ancestors did not have direct access to some infallible measure of the safety, or the fertility, of a particular environment. They did not take soil samples or monitor the crime levels. All they could do was examine a variety of indicators correlated to the fitness of the environment in their experience—experience that valued safety and survival. Similarly, when birds explore potential nesting sites in woodland, they need to be sensitive to a variety of factors concerning the availability of food and security, but ornithologists have discovered that they make their decisions about whether to nest in a particular site on the basis of the abundance and pattern of tree branchings. It is likely that some human choices of suitable habitats were made in response to easily assessible cues, in a similar way. This is a state of affairs that can lead to responses on cue when the primary attribute is not present. Thus, the appearance of clouds on the horizon is a welcome sight in a dusty savannah grassland. Their appearance is strongly correlated with rain and a local abundance of food. Even when you have running water in abundance, a disposition towards finding cloud patterns pleasant would remain as an inherited adaptation, which once had positive survival advantage over a disinterested attitude to the sky.

Psychologists have carried out a number of controlled experiments on children and adults to discover which environments they prefer. By using photographs it is possible to remove extraneous factors (like the presence of water or animals) that are not common to all pictures and to expose the viewers to habitats of which they have had no direct experience. The results are interesting. It was found that among very young children the savannah environment was the most preferred. (The desert was preferred the least.) But older teenagers, who had experienced other environments (like deciduous woodland, rainforests), often liked them just as much as the savannah. The overall pattern of the studies suggests that, among the very young, there is an innate preference for the savannah landscape; this preference is then modified by experience of, and learning about, other environments as the subjects grow older. When experience is limited and the subjects are choosing from photographs of environments of which they lack experience, then the savannah landscape is the most pleasing. There is evidence for an innate bias towards the savannah habitat that, in the absence of overriding experiences of other conditions, creates a natural aesthetic disposition as a legacy of the adaptive success of our early ancestors.

The savannah landscape (Plate 8 and Figure 3.25) is an environment with many reliable cues for safe and fruitful human habitation. These cues are widely reproduced in our parklands and recreation areas. There is scattered tree cover, which offers shade and escape from ferocious predators, interspersed with grasses; yet there are long vistas with frequent undulations that allow good views, orientation, and way-finding. Most food sources are within a metre or two of the ground, whereas in a forested environment life is concentrated, out of reach, high above the ground, and terrestrial creatures are condemned to scavenge for the scraps that fall from the forest canopy.* The most distinctive unpredictability about savannah life is the availability of water. Here, one recognizes the importance of cues like cloud formation, changes in temperature and weather outlook, and seasonal variations in the colour and vitality of plant life, together with the water levels in rivers and streams. Sensitivity to these environment indicators has a clear adaptive advantage over insensitivity. The presence of trees, greenery, and water offers an instant evaluation of the suitability of a potential habitat. These primary indicators, together with a sense of the openness of the terrain, its prospects for shelter, and the furtive viewing of others, are valuable sensitivities that signal whether further exploration or settlement can safely ensue. If the environment is deemed safe for further exploration, then other features highlight the most attractive sites. The topography must allow us to navigate easily; landmarks, bends, and variations are welcome to the eye, so long as they do not create confusing complexities, or mask dangers.

---

* Habitat has a significant effect upon social behaviour. In his influential, though controversial, book *Sociobiology*, E. O. Wilson describes some differences that one might expect to evolve in populations of savannah- and forest-dwellers: 'Forest-dwelling creatures will usually be more solitary than savannah-dwelling ones, who tend to be gregarious: in open spaces there is safety in numbers (for prey); in the forest it is easier to hide if you are solitary and also easier to sneak up on a victim. Solitary animals tend to be more unfriendly (aggressive) to other members of their species, and develop behaviours, such as special displays, whose ultimate effect is to give each individual his own space or territory.' Whereas fast movement and keen eyesight will be favoured in savannah-dwellers, acute senses of smell and hearing will be more advantageous to cryptic forest-dwellers.

3.25 Some natural and man-made savannah-like landscapes: (i) Richmond Park, London; (ii) Woburn Farm, Surrey, engraving, 1759, by Luke Sullivan; (iii) Holkham Hall, Norfolk, drawing (*c.* 1738) of proposed planting of the north lawns by William Kent.

We recognize, also, the encouragement to exploration that is created by the mysterious element in the terrain: the path that leads out of sight or behind a hill. Its further exploration will be safe only if it combines adventure with automatic caution and an instinct to recoil from danger. This surprising fascination with risk and danger attracts us to all manner of cultural embellishments: from horror stories and roller-coaster rides to paintings of shipwrecks (Plate 9) and disaster movies; it springs from an inherited urge to explore and understand environments as fully as possible from the safest possible vantage point (Figure 3.26). The fact that these hazards are potentially fatal is the reason why a desire to inform oneself more fully about their nature has selective advantage over an attitude of apathetic indifference.

There is a clear adaptive advantage to be gained by choosing environments that offer places of security and clear unimpeded views of the terrain—which allow one to see without being seen—tempered by a mysterious invitation to explore. These combinations remain an innate preference: their attractiveness informs many of our aesthetic preferences, from landscape architecture to painting. Extensive views and cosy inglenooks; daunting castles; the tree-house, the 'Little House on the Prairie'; the mysterious door in the wall of the secret garden: so many of the classically seductive landscape scenes combine symbols of refuge and safety, with the prospect of uninterrupted panoramic views; or the enticement to explore, tempered by verdant pastures and water. These comfortable, pastoral scenes appeal to our instinctive sensibilities because of the selective advantages that such attractions first held for our ancient forebears (Figure 3.21). They figure prominently in our best-appreciated landscape gardening, public parks, and gardens, where they are calculated to aid relaxation and induce feelings of ease and well-being. Distinguished architects, like Frank Lloyd Wright, have laid particular emphasis upon the desirability of creating canopies and refuges within buildings, and often set them in opposition to panoramic vistas, or even cascades of water, in order to heighten the feeling of security that these cosy alcoves create. Sloping ceilings, overhangs, gabling, and porches are all architectural features that accentuate the feeling of refuge from the outside world, while balconies, bays, and picture windows meet our desire for a wide-ranging prospect. The skilful use of trees and water in the design of buildings and gardens can reinforce these features. Their denial in many urban building projects has had consequences that are all too plain to see. Concrete, exposed walkways, innumerable blind corners, greyness and banal predictability, which offer no refuge from everyone else, and buildings that offer no enticement to enter: these abominations have led to depression, crime, and emotional disequilibrium (Figure 3.28). Mike Harding's short guide to modern architecture rekindles those fears that the Psalmist had so blissfully dispelled:

3.26 (i) Some works of art appealing to the urge to explore unknown territory: (i) *Grand Canyon of the Colorado*, from John Wesley Powell's *Exploration of the Colorado River of the West*, 1869; 3.26.

The planner is my shepherd
He maketh me to walk; through dark tunnels
and underpasses he forces me to go.
He maketh concrete canyons tower above me.
By the rivers of traffic he maketh me walk.
He knocketh down all that is good, he maketh straight the curves.
He maketh of the city a wasteland and a car park.

Our aesthetic preferences are a fusion of instinct and experience. We would

3.26 (ii) A. Boens, *The Rocher at the Chateau of Attre, Belgium*, 1825.

expect that, in the absence of experience and special influence, our innate sensitiv-ities for these life-supporting features of natural scenes would remain. Indeed, simple landscapes and still-life scenes are usually preferred by those with no special interest in art. A taste for the avant-garde or the abstract is a fruit of experience overriding instinct. Even then, what appeals in man-made art is the symbolic play, or counterplay, on those same adaptive features that have for so long informed traditional artistic images.

'Time and tide wait for no man', but he who is alert to the precursors of significant environmental change will be best equipped to survive it. Our alert-ness and sensitivity to so many of the transient features of our environment—

(i)

(ii)

3.27 (i)–(ii)

(iii)

3.27 Examples of landscapes displaying images of (i) an open prospect illustrated by J. M. W. Turner's *Petworth Park: Tillington Church in the background*, 1828, Tate Gallery; (ii) a landscape dominated by the image of a refuge, illustrated by *The Bard* by John Martin, 1817, Laing Gallery; and (iii) a balance between images of prospect and refuge, illustrated by C. F. Lessing's *Castle on the Rocks*, 1828.

the lengthening shadows that signal the end of daylight; the darkening clouds or rushing winds that herald cold or storm; the distant horizon that hides the unknown 'over the hills and far away'—all are pointers that once rewarded response and appreciation. Our artistic fascination with sunsets and cloud patterns; our sensitivity to the nuances of light and shadow in the representation of the natural world; the menace of the storm and the tempest: all these instinctive feelings make sense as residues of reactions to changes in the environment that require evaluation and response. Shadow reveals new information about distance and depth; it offers the prospect of more detailed appraisal of the environment. Danger lurks in the shadows; it pays to be especially sensitive to it. Alertness to the sunset (Plate 10) and the shadows that signal the coming of darkness, and the need to change patterns of behaviour in order to ensure warmth and safety, has clear advantage over disinterest. Reaction to the appearance of the Sun when it is far from rising and setting, by contrast, offers far

3.28 An unpleasant urban building that offers no sense of providing entry or refuge.

less of vital importance to organisms. You don't need to know that the Sun is overhead in order to tell that you are getting too hot.

With the darkness comes the importance of fire; flickering flames still fascinate us. The fire was the focus of life after dark, offering warmth and safety, fellowship and light. It inflames strong emotions—positive and negative—by its paradoxical offerings of comfort and danger. This odd mixture of fear and fascination appears elsewhere. Large animals are strangely attractive, yet threatening. Large animals were once both a danger and a ready source of abundant food.

Our instinctive attraction to them, tempered by fear and respect, looks like a remnant of a reaction that increased the likelihood of survival, as compared with a response of total fear and isolation, or one of reckless familiarity. Animals were the key to our ancestors' survival. It is not surprising that instinctive reactions to them evolved and spread. The instinctiveness of those reactions explains the propensity we have for symbolism that uses animals. The dominance of the lion, the soaring freedom of the eagle, the evil serpent, the fleetness of the gazelle—these are some of the symbols that trade upon our environmental history.

For tropical savannah-dwellers, daily changes in light and temperature are regular and rapid, but other critical changes are slow and subtle. The most unpredictable element of the savannah landscape is the seasonal variation of the rainfall. We would therefore expect to find adaptations in humans that display sensitivities to indicators of seasonal change and of imminent rainfall and fruit-fulness. We find emotional responses to the seasonal changes in the colours of leaves and shrubs: people flock to New Hampshire for the fall. We find flowers beautiful, therapeutic, and romantic. What hospital ward would be without them? What more frequent gift for a loved one? What more common still-life subject? And, oh, what effort the horticulturalist expends to produce bigger, brighter blooms for our admiration. Our unusual interest in colourful flowers, and the lengths to which we go to cultivate and arrange them, is impressive. We don't eat flowers, but the appearance of flowers is a useful cue that allows different plant forms to be rapidly identified and distinguished. If no flowers are present, then plants are all green, and can be distinguished only by detailed inspection. Flowers also give information about the ripeness of fruit. Thus, while plants burst into flower for reasons that have nothing to do with our likes or dislikes, the fact that a sensitivity to flowers has a purpose, which is adaptive, provides us with a clue to the origin of what would otherwise be an entirely mysterious fascination.

It has become fashionable to regard human aesthetic preferences as entirely subjective responses to learning and nurture. This now seems barely credible. Our sensitivities and emotional responses have not been created out of nothing. The evaluation of environments was a crucial instinct for our distant ancestors—one upon which their very survival depended. The adaptive responses that we have inherited from them form a basis over which our experience is overlaid. In many manifestations of the visual arts we see clear remains of past imperatives, now overlain with symbol or subverted into opposition, perhaps, but undeniably present in our representations and recreations of natural landscapes. Even where artistic representations are heavy with artificial symbolisms of a religious or romantic kind, one can often find a background resonance with echoes of our innate emotions. The backgrounds to portraits and religious works often contain scenes that combine images of safety, danger, and wide-open space. A balance of these three ingredients can arouse conflicting emotions and ambiguity.

One should appreciate that these ideas about the origins of aesthetic response would be regarded as deeply heretical by many art critics, who like to believe that artistic appreciation is immune from 'scientific' analysis. But consider how we have long appreciated the role of mathematical structures in aesthetics. We use particular shapes or symmetrical patterns when we wish to emphasize these underlying mathematical harmonies. Our knowledge of the behaviour of light, or the perception of colour, which was made possible by the studies of physicists, is exploited to the full to create images that are attractive and pleasing to the eye. One might suspect that our affinity for these geometrical and optical patterns is linked to the ease with which the brain can produce mental models of them, and the extent to which they are instantiated into the natural world in situations where their recognition will be rewarded. These important mathematical and optical aspects of aesthetics must be added to the biological perspective that adaptive evolution provides. It sheds light upon our attraction by symbols in art, and reveals why particular images can so effectively be pressed into service to conjure up emotional responses. Art would not be a universal human activity if there were no universal emotional responses and resonances that it could pluck. If extraterrestrial beings evolved by natural selection, then we would expect that their environment would have presented quite different challenges from our own. They would need to have met those challenges by inheriting instinctive reactions to their environment that had survival value. We might expect that they would also retain heightened emotional responses to those aspects of their environment whose appreciation would be advantageous to their survival. Knowing something of their environment and their range of senses (which would also be adapted to their environmental conditions: light levels, sound levels, visibility, and so forth), we could expect images of safe havens, clear vantage-points, and danger to produce instinctive responses. If they provided us with examples of their artistic creations and preferences, this is how we might begin to interpret and understand them. While their symbols of safety, danger, and panorama might have been so transformed by their social practices as to be now unrecognizable, if traces remained, we would be able to take the first steps towards understanding how their minds worked.

▓▓ Figures in a landscape: the dilemma of computer art

If we begin at once to break the ties that bind us to nature and to devote ourselves purely to combination of pure color and independent form, we shall produce works which are mere geometric design, resembling something like a necktie or a carpet.

Wassily Kandinsky

The sources of our affection for natural landscapes shed light upon our responses to unnatural landscapes. The ubiquity of powerful computer systems has created an explosion of computer graphics that adorns galleries, bedrooms, book-jackets, and postcards. The computer can produce images on request, with colours chosen to order. This technology has led to the creation of computer-generated fractal landscapes (Plate 11), which display striking similarities to natural scenes. Our discussion of human adaptation to appreciate landscape features helps us to understand our responses to computer-generated scenes. We can see how their focus upon the small-scale texture of landscapes excludes any recognition of the importance of mingled symbolic associations of prospect, refuge, and hazard. They are dominated by wide-ranging vistas and horizons, but lack the deliberate inclusion of refuge symbols and inducements to explore. They fail to resonate with our evolutionary adaptation for emotional response to particular landscape symbols. They are not landscapes that we feel drawn to enter. Nevertheless, there is something beguiling about these images: something that is shared by many other examples of computer art. In order to identify something of what it is, we might consider some of the fascinating issues raised by computer-generated images that are presented as works of art.

Computer art threatens to overturn centuries of reverence for the concept of an 'original' work of art. For what is the 'original' of a piece of computer art, when one can run off innumerable identical copies on the laser printer? The original displays the marks of the artist's own hand; it bears the artist's signature; it shows the detailed brush-strokes that he used to fashion it. The photocopy lacks all these personal touches. Some feel this to be a subversive devaluation of the work of artists that ultimately will lessen demand for it. But although the computer artist cannot lay a great premium upon the uniqueness of one of his printouts, he can atone for this by the sheer quantity of work that he can produce. A devaluation of the status of the original work of art might even be welcomed in some quarters. It would prevent the ownership of works of art from being largely an activity for the wealthy, and the acquisition and possession of works of art from being for some people merely a branch of financial invest-ment. There are undoubtedly many who would not like to see such an egalitarian revolution occur. Questions such as these show that computer art is challenging; while it may not (yet) have produced works of beauty surpassing those of human artists, it raises new questions about the nature of art. Herbert Franke sees the long-term effect of this rival world of art as a dramatic revolution in our attitude towards art and what we can hope to draw from it:

The demystification of art is one of the most far-reaching effects of the use of computers in the arts. No sooner is it recognized that the creation of art can be formalized, pro-grammed and subjected to mathematical treatment, than all those secrets that used to enshroud art vanish. Similarly with the reception of art; the description of reality in rational terms inevitably leads away from irrational modes of thought, such as the idea that art causes effects that cannot be described scientifically, or that information is passed on to the public by the artist that could not be expressed in any other way. And so art loses its function as a substitute for faith, which it still fulfils here and there.

The reproducibility of computer art is a consequence of its 'push-button' quality. It seems to be dominated by the technology used in its fabrication. Technology is used today in conventional painting; it provides acrylic paints, airbrushes, and other innovative materials and methods, but these can still be regarded as improvements to traditional tools and techniques that are vehicles for expression, rather than the essence of that expression. Computer art, by contrast, seems totally dependent upon the computer for its presentation. It is a reflection of the state of the art of computer technology, and of the structure of particular impersonal algorithms. The artist Gary Glenn attacks it as the ultimate hands-off activity,

Computer art is devoid of sensation; there is no direct encounter with materials. Traditional materials do not hide what has been done; there are brush strokes, chisel marks . . . There is a record of the artist's gesture and presence. There is an absolute lack of humaneness in computer-generated art. Is there an artist who works solely with computers and solely for esthetic or artistic reasons?

Works of computer art have nevertheless been displayed in the world's most famous galleries. There are journals devoted to their appreciation. Multimillion-dollar movies are built around the special artistic effects that only computers can create. But is it really art? Perhaps it depends upon who you ask—and how. Cliff Pickover, a renowned virtuoso of computer graphics at IBM in New York, invited readers of one of his books to send him their opinions. The result was a classic illustration of the biased sample; he records that 'a majority of those who answered "Is Computer Art Really Art?" by sending me electronic computer mail said "yes". A majority of those who wrote their answers to me using paper letters mailed through the conventional mail system, said "no".' It is also hard to evaluate audience response to computer artworks, because many who like computer art find themselves liking it out of admiration for the technical skill that they witness—they like the computer representation of a scene that would be of little or no interest to them aesthetically if it were presented as a painting or a photograph. In the case of intricate patterns, like the Mandelbrot set, the testimony of mathematicians is misleading because their judgement is skewed by an understanding of the remarkable structure that is being represented—which is all the more remarkable because its most extraordinary properties cannot be captured by any finite picture.

Robert Mueller maintains that the images that arise from the implementation of mathematical formulae and computer algorithms fail to be truly artistic because they are essentially secondary: they are representations that are constrained by some external rules.

Though we can say that mathematics is not art, some mathematicians think of themselves as artists of pure form. It seems clear, however, that their elegant and near aesthetic forms fail as art, because they are secondary visual ideas, the product of an intellectual set of restraints, rather than the cause of a felt insight realized in and through visual form.

Mueller feels that, whereas the artist creates images freely, the computer artist is merely exploring the limits of a procedure, or of an algorithm, or of the number of colours that his printer can display. Yet, perhaps the situation is more subtle. The artist may feel untrammelled by technical constraints but, as we have seen, there are unnoticed biases and constraints imposed by our evolutionary history. What we create because of emotional affinity; what we create in order to override that emotional affinity: both are products of constraints, whose influence can be far more overwhelming than those controlling the computer algorist. Still, the reaction of most artists to computer graphics is coloured by the fact that painting is the art-form least encumbered by technological props. Natural pigments spread by bundles of hairs have been subject to very few innovations. An interesting contrast is provided by music. Like painting, it is universal among human cultures, and can be traced back to the dawn of recorded history. But, unlike painting, it has an equally ancient tradition of using artefacts to generate sounds that humans cannot produce naturally. Moreover, in modern times, the production and recording of music has incorporated many kinds of electronic gadgetry. As a consequence, the creation of electronic music is a smaller step from traditional music than computer art is from human art. The distinction between computer-generated music and 'human' music is far less evident to the casual listener than is the distinction between computer art and human art to the casual observer.

Let us return to the question of what attracts us to computer-generated fractal images. We have dwelt upon those byproducts of our evolutionary history that adapted us for survival in early savannah environments, but to reach that stage of development many other, more basic, responses had been honed by natural selection. Perhaps the most basic of all is an ability to sense and classify patterns. This ability enables danger to be identified in the environment, past threats and opportunities to be recognized when they reappear, and patterns of events and collections of things to be classified. It is adaptive to seek out experiences that aid the process of classifying patterns in the environment. There is a broad group of structures, which we class as symmetrical, beautiful, or aesthetic, whose patterns are easy for us to grasp, and to which we might therefore expect to become disposed. Furthermore, we see that living things tend to be distinguished from non-living natural things (as opposed to the manufactured objects that now surround us) by their symmetry. As we have already discussed, living things possess right–left symmetry about a vertical plane; if they move then they do not possess front–back symmetry; and gravity dictates an up–down asymmetry. Any disposition towards detecting, and responding to, patterns with right–left symmetry might turn out to be highly adaptive. It would reveal when another animal was facing in your direction, looking at you. This might be a signal to escape, to prepare for dinner, or to consider the prospect of a possible mate. A response to symmetry will not always be correct, of course; one might be looking at a beautifully rounded rock, rather than a predator. It requires follow-up responses to elicit further information. But the costs of building in a simple instinctive response to symmetry are rather small compared with the benefits. The survival value of rapid pattern-recognition is considerable.

If we can identify patterns in a landscape, then we are more likely to explore it. Again, as with our innate responses to landscape, we are not shackled by these dispositions. They can be overwritten by experience but, in the absence of individual formative experiences, our inherited responses to patterns will be the default response. And, as with other activities with high survival value, like eating, or returning home safely, they will inevitably become pleasurable. In the case of fractal patterns, we are exposed to a highly developed form of organized pattern that is also present in the natural world (in leaves, trees, and rock formations); it is therefore not surprising that our ability to identify, sort, and classify patterns is activated and engrossed by fractal works of art. But the blandness and uninviting character of fractal landscapes witness to their inability to excite the more habitat-specific responses that attractive natural landscapes evoke. All computer art is heavily biased towards attracting the attention of our brain's most basic pattern-recognition skills, and the fact that this form of representation tends to exclude most traditional forms of symbolism only serves to accentuate the response to patterns. There is ample scope for aesthetic appreciation to blossom as a byproduct of selection for pattern-recognition. By regarding pattern-recognition as a type of game played against a potential environmental threat, we can see why we might expect our minds to be over-sensitive to the presence of patterns. The negative consequences of 'seeing' patterns in the undergrowth when there is no lion lurking there are very small compared with the fatal consequences of failing to identify a lion when it *is* there. A tendency towards paranoia, self-deception, and an over-sensitivity to the presence of patterns is thus understandable.

This sensitivity to identifying patterns has manifestations that are especially interesting because they appear in many Middle Eastern cultures where the artistic representation of living things is forbidden. The more one looks at the ordered renderings of computer art, with their emphasis upon symmetry and reflection, the more one feels that they are examples of pattern exploration, rather than of art. One can imagine an ancient debate between Arabs and Euro-peans as to whether their respective forms of art were 'really' art. Throughout history, humans have produced decorative designs in the form of mosiacs, tilings, and friezes. The Islamic tradition is particularly notable in this respect, because the teachings of the Koran forbid the representation of living things for decorative purposes. The Arabs consequently exploited the whole spectrum of complexity that geometry allows, both on flat and curved surfaces. The geo-metrical intuition of their artists surpassed that displayed by contemporary mathematicians. They have much in common with Maurits Escher's work, which has also stimulated new discoveries in geometry (Figure 3.29).

In these examples one sees our instincts for the recognition, generation, and classification of patterns at work. The most widespread systematic use of decora-tive patterns is the simplest: the creation of linear friezes. The range of alterna-tives is not as great as the wallpaper catalogues would lead one to believe. There are only seven linear patterns which can be repeated on a strip of paper to produce a frieze using two colours; the total number of repeating patterns that can be created on a plane surface to create a frieze, using only two colours, is seventeen.*

When two colours (say, black and white) are used to produce a linear frieze, there are only four basic ingredients that can be employed to create a repeating pattern. The first is *translation*: just moving a pattern along the frieze, *en bloc*. The second is *reflection* about a vertical or horizontal axis. The third is *rotation* through 180 degrees around a fixed point. The fourth is a *glide reflection*, which consists of a forward translation together with a reflection of the image about a line parallel to the direction in which it is translated, and results in the mirror images created by the reflection being slightly offset from one another, rather than vertically aligned. Each of these four moves is shown in Figure 3.30. These four operations can be combined in only seven different ways to produce re-peating designs as shown in Figure 3.31. The different possibilities arise by acting upon some initial motif, which need have no symmetry, with the following operations:

(a) translation
(b) horizontal reflection
(c) glide reflection
(d) vertical reflection
(e) rotation through 180 degrees
(f) horizontal/vertical reflection
(g) rotation/vertical reflection.

* If there are C colours then the number of different patterns is 7 when C is an odd number, 17 when C divided by 4 leaves a remainder of 2, and 19 when C is exactly divisible by 4.

(i)

(ii)

3.29 (i) Several Islamic tiling patterns used by Moorish designers of the Alhambra, sketched by Maurits Escher; (ii) *Eight Heads*, Maurits Escher, 1922. An early woodcut made when the artist was a pupil at the School of Architecture and Decorative Arts in Haarlem.

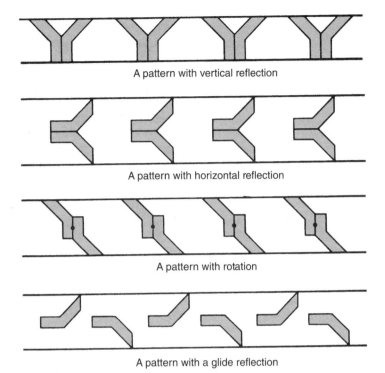

A pattern with vertical reflection

A pattern with horizontal reflection

A pattern with rotation

A pattern with a glide reflection

3.30 The four basic operations that can be used to generate a frieze.

Examples of the seven possible varieties of frieze patterns are found in decorations all over the ancient world: from the pottery of San Ildefonso, to the vases of the Incas, and traditional forms of Maori decoration. Some magnificent examples of the seven, taken from a diverse selection of cultures, are shown in Figure 3.32.

Let us move up a dimension from friezes to wallpaper. Symmetrical patterns in two dimensions have more freedom to reproduce using combinations of the basic reflections, translations, and rotations. There are seventeen possibilities, which were first classified by Eugraf Federov in 1881; but it appears that they were all known, and employed for decorative purposes, by the ancient Egyptians. The most spectacular renderings of them are to be found in the Moorish decoration of the Alhambra (see Figure 3.29(i)). The seventeen examples are displayed in Figure 3.33, using examples gathered from ancient decorations in a wide range of cultures. If one departs from these regular designs, which all have a lattice-like structure that is invariant after traversing vertical or horizontal directions, then the number of possible designs grows dramatically. In fact, any one of the patterns can then be combined with the others in an infinite number of different permutations.

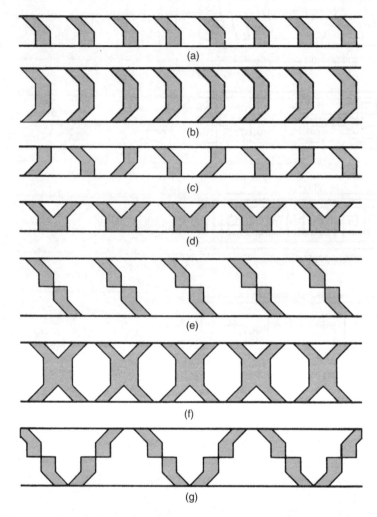

3.31 The seven distinct friezes that can be generated by combinations of the four basic operations shown in Figure 3.30. The labels correspond to the combinations of these operations listed on page 127 of the text.

The ubiquity of these forms of decoration, in cultures with no mathematical understanding of their significance and completeness, witnesses to the innate human sensitivity towards patterns—a sensitivity that has clear adaptive advantages. In the ancient world the equivalent of the modern contrast between computer art, landscape, and other forms of representational art was to be found in the contrast between decoration and the representation of living things and environments. The enduring attraction of both types of image witnesses to the different threads in our patchwork of aesthetic appreciations. In the most

Dragon and phoenix carpet, Asia Minor (I) Chinese ornament painted on porcelain

Masonry fret, temple at Milta, Mexico (II) Stained glass, Cathedral of Bourges

Greek fret with a vase (III) French Renaissance ornament from casket

Greek fret (IV) Ancient Greek scroll border

Pompeian mosaic (V) Maltese lace

Chinese ornament painted on porcelein (VI) Indian painted lacquer work

Modern rug (VII) Italian damask of the Renaissance

3.32 The seven possible frieze symmetries, each illustrated by two examples from the decorative traditions of different cultures.

(I) Egyptian wool fabric

(II) French medieval wall ornament

American Indian industry

Ghordes rug

(III) Akhal Tekka prayer rug

Medieval illuminated manuscript

Egyptian mat design

Shiraz rug

(V) Italian velvet of the Renaissance

Decorations from spandrels of arches, Alhambra

Mosaic pavement, Baptistery, Florence

French Renaissance carpet

(VII) Sixteenth-century textile

Sixteenth-century textile

German Renaissance embroidery (Bavaria)

(VIII) Fourteenth-century moorish silk

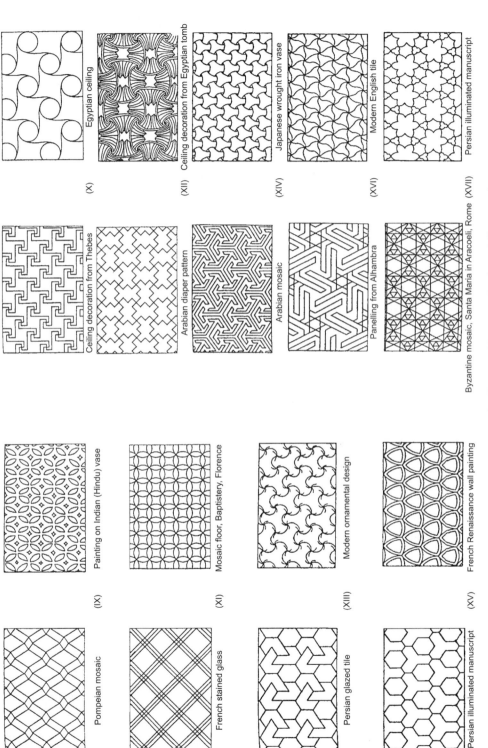

3.33 The seventeen possible two-dimensional patterns ('wallpapers'), illustrated by examples from different cultures.

Pompeian mosaic

Painting on Indian (Hindu) vase (IX)

Ceiling decoration from Thebes

Egyptian ceiling (X)

French stained glass

Mosaic floor, Baptistery, Florence (XI)

Arabian diaper pattern

Ceiling decoration from Egyptian tomb (XII)

Persian glazed tile

Modern ornamental design (XIII)

Arabian mosaic

Japanese wrought iron vase (XIV)

Persian illuminated manuscript

French Renaissance wall painting (XV)

Panelling from Alhambra

Modern English tile (XVI)

Byzantine mosaic, Santa Maria in Aracoeli, Rome (XVII)

Persian illuminated manuscript

traditional forms of painting the symbolic resonances are dominant over the instinctive recognition of pattern but, as we shall see in a later chapter, the roles are reversed in our appreciation of patterns of sound.

Midnight's children: a first glimpse of the stars

> The sensuous contrast of the dark background,—blacker the clearer the night and the more stars we can see,—with the palpitating fire of the stars themselves, could not be exceeded by any possible device.
>
> George Santayana

This chapter began with the stars. From them, the biochemical building blocks of complexity emerged, together with the rays of heat and light that promote and sustain the novel form of complexity that we call life. We have seen how invariant features of the fabric of the Universe fix the sizes of the heavenly bodies, and planets, in ways that constrain the forms and sizes of structures and organisms on the Earth's surface. Size, it appears, is all-pervasive in its influence, touching the range and time-scales of life in unsuspected ways. As those constraints of natural environment were accommodated by the evolution of adaptive organisms, selection led to curious sensitivities to the environment, whose legacies are manifest in our aesthetic feelings for natural scenes, and in many of our antipathies to the unnatural. These considerations reveal something of our intuitions for the natural and the unnatural; they lie at the heart of our latent desires to appreciate, nurture, and recreate the environment in sensitive ways. They teach us something about our responses to symbols, and about the pleasure we derive from symmetrical patterns. And so what a pleasurable conclusion it would make if in our end could be our beginning—if we could return to the stars to close this part of our story. Alas, it is not to be. We have inherited no emotional responses to the stars from our ancestors; there is no tradition of painting the night sky. Nor would we expect it. The heavens change slowly. Whereas the sunset stimulates the emotional responses needed to accommodate rapidly changing circumstances, the coming of the stars signals nothing so urgent in the mind of the hunter or the gatherer. The heavenly tapestry is an acquired taste; but its influence, although subtle, is no less pervasive. As we are about to discover.

4 | The heavens and the Earth

Science is spectrum analysis.
Art is photosynthesis.

KARL KRAUSS

The remains of the day: rhythms of life

Education is an admirable thing, but it is well to remember from time to time that nothing that is worth knowing can be taught.

Oscar Wilde

Every month, faceless organizations send me bills. Every quarter, others join them. And, as the New Year begins, another coterie of computers resolves to display my address through the windows of manila envelopes. These periodic communications are repeated the world over, and the labyrinthine pattern of our daily lives is held together by a skeleton of days, months, and years. By these cycles, we structure time and organize our lives, and reflect the celestial patterns that have stimulated and constrained the evolution of our environment. Days and nights, seasons and tides, cycles of fertility, rest and activity: all are reflections of the rhythms imposed upon us by celestial motions. They have influenced where, and how, people may live; the elements that they must overcome; the shelter and dress they must construct, and the stories that they tell about it all. By these devices and desires, the inexorable motions of the heavens and the Earth have cast their shadows upon our bodies, our actions, and our superstitions about the meaning of the world. In this chapter we shall explore some of the unexpected connections between the heavenly bodies and the pattern of life on Earth. We shall look at these links at different levels, beginning with the underlying temporal patterns in the terrestrial environment, and ending with the learned human responses to the astronomical realm. These responses still manifest themselves through our social organization and they also underlie many of our metaphysical and emotional responses to the Universe. We have been tempted to see stars as gods, as demons, as navigational guides, as omens of bad luck, or, at worst, as the rulers of our every action. We shall also discover that we have been unusually fortunate to find ourselves living, by chance, in celestial circumstances that influence significantly the scope and direction of any scientific investigation of the Universe. By appreciating some of the delicacies of this situation, we shall be better placed to evaluate the likelihood of extraterrestrial organisms attaining the level of scientific understanding of the Universe that we have achieved. We shall see that progress is not just a matter of intelligence; it depends greatly upon one's vantage point in the Universe.

Our early ancestors' first preconscious steps along the evolutionary pathway were taken in a world with a daily alternation of light and day, a monthly ebb and flow of the tides, and an annual variation in daylight hours and of climate. All these changes of scene have left their imprint upon the actors in the serial drama of life. Some survived best because chance variations gave them body rhythms that closely mirror the pulse of advantageous changes in the environment. Others acutely sensed some aspect of the rhythms of the heavens directly, and responded to their marching orders. The world is full of plants and animals that have grown sensitive to the cycle of night and day, the seasonal cycle of the Sun's heat, and the monthly pull of the tides. Ocean tides raised by the waxing and waning of the Moon influenced the evolution of crustaceans and amphibians. The development of intertidal regions in which conditions alternated between submersion and drying may have encouraged the spread of life from sea to land. Changing conditions stimulate the evolution of a breed of complexity that leads ultimately to life because it creates conditions in which variation makes a difference to the prospects for survival.

There are clear imprints of an annual period in life-cycles of animals. Evolutionary adaptation will favour the survival of innate 'clocks' that time the birth of offspring to coincide with times when the chances of survival are highest, especially in the temperate regions where the seasons change abruptly. An impressive example is provided by the spawning of the grunion fish in southern Californian waters. They spawn at the highest reach of the spring tide, when the Moon is dark or full, leaving their spawn after burrowing half of their bodies into the sand. As successive tides are lower, the eggs remain out of reach of marine predators. They hatch two weeks later, when the tide has turned, just in time to be helped into the sea by the next advancing high tide. A lack of respect for this tidal cycle would be penalized by predators, and organisms with innate timing-triggers in step with tidal variations will prosper at the expense of those that lack them. Because tidal forces are manifestations of the same monthly cycle of lunar variations that alter the fraction of the Moon's face that can be seen by reflected sunlight at night, it is possible to synchronize with tidal cycles by various means: by sensing the forces directly, by sensing moonlight variations, or by behavioural variations in the intertidal region.

Animals sense the changing of the seasons by a response to the length of the hours of daylight. There are remarkable examples of the accuracy of this sensing, which optimizes female fertility to coincide with the spring equinox. A critical daylight length seems to trigger mating activity. Experiments show there may be just two phases: light-loving and dark-loving. In the first phase, when light falls on the body it enhances growth and activity; in the second phase these things are inhibited. On long days, more light stimulates stronger biochemical responses. Yet the situation is not always so simple. Creatures can have their internal clocks reset by exposing them to artificial environments. Much argument has occurred among biologists about the respective roles of internal, genetically regulated clocks and of external influences in explaining biological cycles. It appears that living things have baseline rhythms, inherited through adaptations to the environment, which can be shifted by changes in the environment and entrained into new cycles.

The day and the year are the simplest of our time divisions. The length of the day is determined by the period taken by the Earth to spin round once upon its axis. The day would last much longer if the Earth rotated more slowly, and diurnal variations would not exist at all if the Earth possessed no rotation. In that case, living things would be divided into three distinct populations: one for the dark side, one for the light side, and a third for the twilight zone in between. The day could not be dramatically shorter because there is a limit to how fast a body can spin before it starts to part company with things on its surface and disintegrate. The length of the day is in fact very slowly lengthening, by about two-thousandths of a second every century, because of the pull of the Moon. Over the vast periods of time required for significant geological or biological change, this small increase becomes quite significant. The day would have been eleven hours shorter two thousand million years ago when the oldest known fossilized bacteria were alive. Direct evidence of this change imprinting itself upon living things has been found in some coral reefs in the Bahamas. Daily and annual growth bands (rather like tree rings) are laid down in the coral, and by counting how many daily bands are in each annual band one can determine how many daily cycles there were in a year. Contemporary coral growths display about 365 bands for each year, roughly as expected, while 350 million-year-old corals, nearby, display about 400 daily rings in each annual band, indicating that the day was then only about 21.9 hours long. This is almost exactly the value that we would expect at that time in the past, given the rate at which the Moon's pull is changing. If we extrapolate back to the formation of the Earth, then the young Earth might have had days lasting only about six hours. Thus, if the Moon did not exist, our day would probably be only a quarter of its present length. This would have consequences for the Earth's magnetic field as well. With a day of only six hours, the more rapid rotation of charged particles within the Earth would produce a terrestrial field about three times stronger than at present.

Magnetic sensing would be a more cost-effective adaptation for living things on such a world. But the most far-reaching environmental effects of a shorter day would follow from the far stronger winds that would whip across the planet's rotating surface. The extent of erosion by wind and waves would be very great. There would be selective pressure towards smaller trees, and for plants to grow smaller, stronger leaves that were less susceptible to removal. This might well alter the course of the evolution of the Earth's atmosphere by delaying the early conversion of its carbon dioxide atmosphere into oxygen by the action of photosynthesis.

The year is determined by the time that it takes for the Earth to complete one orbit of the Sun. This period of time is by no means haphazard. The temperatures and energy output from stable stars are fixed by the unchanging strengths of the forces of Nature. Biological activity can occur on a planet only if its surface temperature is not extreme. Too hot, and molecules fry; too cold, and they freeze; but, in between, there is a range in which they can multiply, and grow in complexity. The narrow range within which water is liquid may well be the optimal one for the spontaneous evolution of life. Water offers a wonderful environment for the evolution of complex chemistry because it enhances both the mobility and the build-up of large concentrations of molecules.

These constraints of temperature ensure that living beings must find themselves on planets that are neither too close, nor too far, from the star they orbit. They will lie in a 'habitable zone' around a central star of the middle-aged sort that is typified by the Sun. Those orbits will need to be quite close to circular if these planets are to stay in the habitable zone throughout their orbital journeys. If they move in wildly eccentric oval orbits, like those of the comets that periodically pass our way, they will then alternately experience conditions of extreme cold and intense heat, rendering the evolution of complexity and life most unlikely. The law of gravitation fixes the time that a planet will take to complete its orbit if its distance from the parent star is known. Planets that are habitable thus have the length of their 'year' determined very closely by unalterable constants of Nature.

These considerations show us that planet-based life will find itself in a periodic environment. Moreover, the cycles of change introduced by its rotation, and by its motion round its parent star, will be not dissimilar to those that characterize our own situation, because all are strongly linked to the conditions necessary for the maintenance of any constant habitable environment. Adaptations to periodic change will be ones that all intelligent life should share.*

---

* If we look around our own solar system, we find that the planets have 'days' varying in length from about ten hours (Jupiter and Saturn) to about 243 Earth-days (Venus); and 'years' between one-sixth (Mercury) and 248 Earth-years (Pluto); see Table 4.1.

One can speculate about which aspects of the world would have left the deepest imprint upon our common view of the world in primitive antiquity. There is the clear division between the Earth and sky, separated by the horizon; the pull of the Earth's gravity orients 'up' and 'down', wherever we go. These experiences are invariable; but others, like the cycles of darkness and light, are periodic. The Sun dominates the daytime hours—the source of heat and light. At night, its role is taken by the Moon and the stars, which straddle the sky in the fuzzy band that we call the Milky Way. All conscious beings on habitable planets orbiting stable stars will be under similar influences. Sun-gods and moon-gods are the most widespread objects of worship in human history; their veneration may well extend far beyond the bounds of our solar system.

▓ Empire of the Sun: the reasons for the seasons

I read, much of the night, and go south in the winter.

T. S. Eliot, *The Waste Land*

As the Earth makes its annual circuit of the Sun, it traces out an orbit that is elliptical in shape. Its greatest distance from the Sun is 1.017 times the average, and its least distance is just 0.983 times the average. This slight deviation from a perfect circle produces an annual variation of about 7 per cent in the flux of energy that the Earth's surface receives from the Sun. The closeness of the Earth's orbit to a circle is clearly important. For Mars, the variation in solar heating is a staggering 90 per cent. Such dramatic variations present significant challenges to the adaptive powers of organisms.

Despite what most people expect, the small annual variation in the distance of the Earth from the Sun has little or nothing to do with the seasonal changes in the Earth's climate. How could it, when Australian summers coincide with European winters? If we divide the Earth's elliptical orbit into four quadrants, we can see that, because it spends more time in the quadrants in which it is farther from the Sun, it in fact receives an equal flux of solar energy while traversing each of the four quadrants. This is a consequence of the inverse square laws of gravitation and illumination.

The key to the Earth's seasonal variations, and to all the diversity that flows from them, is a little accident of its formation: the fact that its axis of rotation is tilted with respect to the plane in which it orbits the Sun. If you imagine the Earth orbiting the Sun on the surface of a table then the table-top specifies the plane of the Earth's orbit. This plane is called the *ecliptic*. As the Earth orbits the Sun it rotates round its Polar axis every 23 hours and 56 minutes; but the Polar axis is not perpendicular to the ecliptic: it is obliquely inclined with respect to it at an angle of 23.5 degrees.

It is this modest tilt that makes the Earth's surface such a diverse place. The Earth maintains its orientation relative to the distant stars as it orbits the Sun, and so its obliquity ensures that different hemispheres receive different fluxes of solar energy. Two lines of latitude, known as the Tropics of Cancer and Capricorn, have latitudes equal to 23.5 degrees north and south respectively; within those latitudes, the length of daylight hardly varies, and there is a day of the year when the Sun is directly overhead. By contrast, within the two Polar circles located at latitudes 66.5 degrees north and south, there are huge variations in daylight hours: the Sun does not rise at all for part of the winter, and does not set for part of the summer ('the land of the midnight sun'). In the temperate zones, between the Tropics and the Polar circles, the Sun passes much higher in the sky during the summer than in the winter; in consequence, summer daylight hours are significantly longer, and temperatures are higher (Figure 4.1). By contrast, in the Tropics there is little variation of temperature between the seasons. Rather, they are characterized by an alternation of wet and dry periods, with their variations in plant and insect life, and by the associated diseases that follow the changes in humidity.

The Earth would have been a far duller place if its rotation axis was not tilted away from the perpendicular to the orbital plane. This, together with other properties of the Earth, can be seen in the context of the other planets in Table 4.1. If there was no tilt, there would be no seasons. The Sun would rise each morning, and set each evening, after following the same daily path through the sky. The hours of darkness and daylight would be equal everywhere; climate would be steady; winds more moderate and, without seasons, climatic zones would be sharply defined by latitude alone. The flora and fauna would be very specialized because each species would occupy particular unchanging environments. In the last chapter, we saw how climate can influence the sizes of living things. As a result, on Earth there are significant trends in the size and diversity

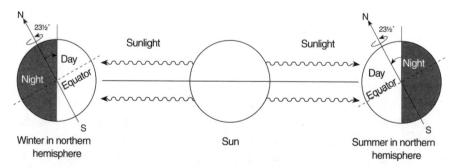

4.1 The reason for the seasons: the tilt of the Earth's rotation axis with respect to the plane of the Earth's orbit around the Sun (not drawn to scale).

Table 4.1 Some data on the planets and the Moon. 'Days' are Earth solar days and 'years' are Earth years. The giant outer planets (Jupiter, Saturn, Uranus, and Neptune) are fluid down to significant depths and have no well-defined surface. The values of their surface temperatures, surface gravities, and compositions are specified for a layer of the atmosphere where the pressure equals that of the Earth's atmosphere at sea-level.

	Mercury	Venus	Earth	Moon	Mars	Jupiter	Saturn	Uranus	Neptune	Pluto
Distance from the sun (10^6 km)	57.9	108.2	149.6	149.6	227.9	778.3	1427	2870	4497	5900
Orbital period	88 days	224.7 days	365.3 days	27.3 days	687 days	11.87 years	29.46 years	84.01 years	164.79 years	247.69 years
Axial rotation period (days)	58.64	243.0 retrograde	0.997	27.32	1.026	0.410	0.444	0.718 retrograde	0.768	6.387
Tilt of rotation axis	$0.0°$	$177.3°$	$23.5°$	$6.7°$	$24.0°$	$3.1°$	$26.7°$	$97.9°$	$29.6°$	$94°$
Radius at equator (10^3 km)	2.439	6.052	6.378	1.738	3.394	71.4	60.0	25.6	24.3	1.18
Mass (10^{24} kg)	0.330	4.87	5.98	0.0735	0.641	1900	569	86.8	102	0.014
Gravity at surface (metres per sec^2)	3.7	8.9	9.8	1.6	3.7	24.8	10.6	8.9	11.6	0.4
Average surface temperature (°C)	170	460	15	1	-50	-143	-195	-201	-220	-205 to -165
Main components of atmosphere	He, Na, O	CO_2	N_2, O_2	Ne, Ar, H_2, He	CO_2	H_2, He, CH_4	H_2, He, CH_4	H_2, He	H_2, He	CH_4, ?N_2, ?CO

of living things as one moves away from the stable environment of the Equatorial regions towards the vagaries of the Polar extremes. All this variation is a consequence of the tilt of the Earth's axis of rotation. Without it, the intermingling of creatures of different sizes would have no climatic restriction, and the Earth's ecology would be very different.

By contrast, if the Earth's axis were tilted much more than it is, then conditions would become far more hostile. The most extreme seasonal variations would occur with a tilt of 90 degrees, in which case the Earth's rotation axis would lie in the plane of orbital rotation.* Seasonal changes would be far more abrupt and extreme. The Earth's surface would oscillate sharply between tropical summers and arctic winters. Extensive ice-caps would form and melt each year, leading to huge variations in sea-level. If the Earth's tilt were 90 degrees, then the melting of polar ice would produce increases in sea-level exceeding 30 metres every six months. Continental land-masses would be reduced in size and the area of the planet available for life to evolve would fall significantly. Life-forms would need to be extremely mobile to cope with the stark seasonal changes. All animals would need larger ranges, and would be much more susceptible to extinction if sudden geological changes prevented them from migrating to warmer climes. Wind speeds would be far higher; storms stronger, and more prevalent. The Polar circles would encompass more of the Earth's surface, and small animals would find their habitats reduced and more crowded by competitors. In effect, the Earth would be a smaller, less hospitable place for living things. Far less of its surface would remain for long periods at a temperature conducive to life; and far less still would be within the range of seasonal variation that the process of evolutionary adaptation could keep pace with.

In Victorian times, it was fashionable for scientists and theologians of a certain persuasion to produce apologetic works, which set forth a wondrous collection of features displayed by Nature, without which human life would be intolerable, if not impossible. These features of the world were invariably presented as convincing evidence for benevolent design at the heart of Nature—with ourselves as the principal beneficiary. The life-supporting features of the natural world were displayed as being so unusual and essential that they could have arisen only by the purposeful intent of some Grand Designer. Hence, an argument for the existence of God was supported by highlighting the miraculous abundance of circumstances conducive to the maintenance of human life. This style of argument spawned an entire subdiscipline of 'natural theology',

---

* This is similar to the situation for the planet Uranus, whose axis of rotation is tilted by 98 degrees towards its orbital plane. This extreme situation may be the result of an impact by another body soon after the solar system formed.

which was especially prevalent in England, and attracted the support of many famous scientists. Not surprisingly, the tilt of the Earth's axis of rotation relative to its orbital plane was one of the features emphasized by supporters of these Design Arguments. We should stress that we are not seeking to reiterate arguments such as these. Our logic is oppositely directed. Rather than infer anything metaphysical from the form of the celestial motions, or conclude that they have been set up to allow life to exist, we wish to show how the celestial arrangements have inevitably influenced the forms of life that arise, evolve, and spread on the Earth. Although changes in some of these features of the solar system would render life on Earth impossible (especially if the changes were large), others would not. Life would still arise in these changed circumstances, and would display the adaptations appropriate to them.*

The closeness of Earth's orbit to circularity means that its tilt dominates the annual variations in climate. If the orbit were far from circularity, this would no longer be the case. An interesting example of this sort is Mars, which has a rotation period that makes its day of similar duration to the Earth's (24 hours 37 minutes). Its rotation axis is tilted by 24 degrees, an angle very similar to that of the Earth (although it varies between 16 and 35 degrees over a period of 160 000 years). The climatic variation on Mars is, however, dramatically greater than on the Earth, simply because it is dominated by the variation in solar energy it receives throughout the long Martian 'year'. Moreover, without oceans to act as a sink for these changes in temperature, and with huge variations in surface topography, its climatic variations are extreme.

The Earth's degree of tilt is a happy medium. We cannot conclude, like the natural theologians of old, that this tilt is optimal—that we live 'in the best of all possible worlds'—or that life could not have evolved on Earth if its tilt were significantly different (although that might well be true). Instead, we illustrate how the rhythm of the seasons and the climatic variations on the Earth, which have fashioned so many avenues of human and animal development, bear the imprint of the structure of the solar system.

---

* Recently, a very speculative idea has been put forward by the cosmologist Edward Harrison of the University of Massachusetts. Cosmological theories of the very early universe have revealed that it is possible, in principle, to create within a microscopically small region of space the conditions required to make that space expand at a speed close to that of light and produce an astonomically large region whose inhabitants would refer to it as the 'observable universe'. Although this capability is far beyond even the dreams of our current technology, it is not inconceivable that a highly advanced scientific civilization might have this capability. If so, Harrison speculates that they will be in a position to determine the local conditions that exist in the regions that they make expand dramatically. In fact, it seems that they can also influence the effective values of some of the constants of Nature which define their environment. Hence, an advanced civilization could deliberately 'tune' conditions to be life-supporting in future generations of large, expanding mini-universes. They would take their cue from the fortuitous conditions they first discovered underwriting their own existence.

◼ Extrasolar planets: a case of spatial prejudice

> God help me in my search for truth, and protect me from those who
> believe they have found it.
>
> Old English Prayer

Up until 1995 we could do nothing more than wonder if the odd properties of our solar system that have made life possible and fashioned our perspective on the Universe were fortuitous or not. They certainly seemed special, but with a sample of one there was not much more to say. It might have been that planetary systems formed only when some rare cataclysmic explosion occurred, or it might be that they arise naturally whenever a star like the Sun forms. Then, rather suddenly, everything changed. On 5 October 1995 Michel Mayor and Didier Queloz of the Geneva Observatory announced the first detection of a planet outside our solar system. It was detected in a 4.23077 Earth-day orbit around the star 51 Pegasi. Its mass was similar to that of the planet Jupiter, a thousand times greater than the Earth. Just one week later Geoff Marcy and Paul Butler, then at San Francisco State University, confirmed its presence and led the race to detect a host of other planets. Today the count stands at over 130, and the discovery of a new one doesn't even make it into the newspapers nowadays unless there is something special about the planet or its orbit.

As the catalogue of extrasolar planets has grown we have learned a number of important and slightly confusing things about planets. It is clear that planet formation is a general process in the Universe. In this respect our own planet is not a special case. But as the number of extrasolar planets built up in the catalogue we began to see the ways in which our solar system is distinctive. Of the known planets, only 14 are in planetary systems of more than one planet. Eleven of those show two planets in orbit and two of them show three. It is important to recognize that this may not be a complete inventory of the planets even in these systems because the observational technique is only sensitive enough to detect giant planets like Jupiter. Astronomers monitor the 'wobble' in the star's position that is created by the orbiting planets. Big planets produce bigger wobbles. These 'jupiters' are great balls of liquid and gaseous hydrogen with no solid surface. They are not places where we will find conventional forms of life. However, they may well possess systems of small moons which are solid like the Earth.

When we look at the orbits of these planets we make the most interesting discovery of all. All the planets that lie very close to their parent star are in almost circular orbits. This is what we would expect. The gravitational forces exerted on the planets by their parent stars force them into circular orbits. What we did not expect, though, was to find giant gaseous planets orbiting so close to their parent star. We don't know how they could have formed so close to their star, so maybe they formed farther out and migrated inwards as the system aged. But this is not the only puzzle. As we move farther out the orbits are found to be extremely eccentric oval shapes. This is in stark contrast to the situation in our solar system where the orbits are almost circular.

This strange state of affairs may be telling us something about what is needed for life to evolve. If planets move in circular orbits, then they feel the same average climatic conditions all the way around their orbital year. But if their orbits are highly eccentric they are going to fry when they come in close to the star and then freeze when the orbit takes them far away. There is dramatic climatic variation around their year, with any surface water freezing and thawing (even boiling) with considerable regularity. This could be too challenging an environment for life to get a foothold on the evolutionary ladder. Around each star there is a habitable zone within which water can exist in liquid form on the surface of an orbiting planet. Circular orbits stay within the habitable zone throughout their orbits; elliptical orbits will generally leave the habitable zone. For some reason our solar system has simple circular orbital motions. One of the orbiting planets in that system sits nicely in the middle of the habitable zone and that planet is where we live. Our discovery of many extrasolar planets have been strangely ambiguous in their message. On the one hand, we are convinced that planets like Jupiter are common and we expect eventually to find that solid planets and moons the size of the Earth are too. But on the other, we have come to appreciate that the *motion* of the Earth around the Sun is special. Eventually we will be able to determine something of the rates of rotation and angles of tilt of Earth-like planets. Again, we will be able to judge how special the terrestrial situation really is by using real evidence rather than mere speculation.

A handful of dust: the Earth below

> The geological formations of the globe already noted are catalogued thus: The Primary, or lower one, consists of rocks, bones of mired mules, gas-pipes, miners' tools, antique statues minus the nose, Spanish doubloons and ancestors. The Secondary is largely made up of red worms and moles. The Tertiary comprises railway tracks, patent pavements, grass, snakes, mouldy boots, beer bottles, tomato cans, intoxicated citizens, garbage, anarchists, snap-dogs and fools.
>
> Ambrose Bierce

The surface geography and subterranean geology of the Earth contribute to its uniqueness in subtle ways that make our own existence and behaviour patterns possible. The arrangement of the continental land-masses relative to the axis of rotation is an interesting example (Figure 4.2). The early spread of humanity's influence after the development of agriculture was more easily accomplished over continents that straddled lines of constant seasonal climate, than over those land-masses that ran across a whole range of climatic variations. Eurasia extends over vast distances, west to east, along lines of constant latitude, whilst the Americas run north–south. Consequently, it is harder for plants and animals to spread down the Americas than across Eurasia, because of the additional adaptation required in order to live in a different climate. A temperate zone runs from Britain across to China, and domesticated animals and cereals are fairly universal across the Eurasian continent. By contrast, the tropical region separating North and South America was sufficient to prevent the migration of animals and crops between them. If lines of constant temperature, or the orientations of the continental land-masses, were rotated through 90 degrees, then the early settlement and development of the Americas would have been quite different. The rise of agriculture in the New World would have been faster, and its civilizations would have matured and spread more rapidly than those of the Old World. Thus, geography and astronomy set the stage for the evolution of life and culture. The spread of plants and animals is followed by their cultivators and husbanders. With them comes language and custom, trade and influence.

The internal composition of the Earth also has profound implications for us. All our fuels are fossilized gases, liquids, and solids, extracted from beneath its surface. Oil and gas build up in places where a porous layer of rock has been laid

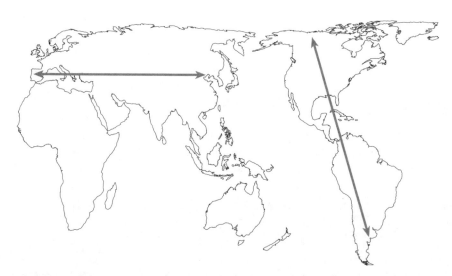

4.2 The orientation of the continents in recent geological history has expedited the migration of cereals across Eurasia because it is spanned by zones of similar latitude and climate. The opposite is the case in the Americas.

down in a particular configuration under an impermeable layer. There is, unfortunately, no way of predicting where these deposits are situated by merely inspecting the Earth's surface. It is much the same for metals and other useful minerals: easily accessible surface deposits were exhausted long ago, and deep underground searches are necessary to locate new reserves. If we fail to locate further supplies of particular metals and minerals in the future, then industrial societies will gradually fade away in the absence of fuels and raw materials for building and manufacturing. Again, when we come to speculate about the likelihood of extraterrestrials, this gives us pause for thought. The development of advanced technology and science by the inhabitants of a planet demands huge supplies of metallic ores and other special materials. The presence of these materials also influences the course of scientific development. For example, the strength of a planet's magnetic field will determine how vital it is for the planet's inhabitants to understand the phenomenon of magnetism early in their development; if there are seas covering most of the planet's surface in between habitable land-masses, then the study of astronomy will become vital for navigation.

The production of concentrations of the heavy metallic ores that are so technically useful may require a rather special state of affairs to exist on the planet: a state of affairs that, in the solar system, is unique to the Earth. The existence of long-lived movements of the Earth's interior, and a cycle of erosion to transport soluble metallic compounds by the global movement of water, plays a key role in this process. The Earth's surface is divided into several fairly rigid areas, termed 'plates'; there is very little movement within the body of a plate, but movements at the plate boundaries are common and have dramatic consequences: earthquakes, volcanoes, new mountain ranges, and ocean trenches.

The Earth's surface possesses a number of simple but profound features, without which the development of life would have been inhibited, or prevented. The division of the Earth's surface between water (70 per cent) and dry land (30 per cent) has played a key role in dictating the directions in which evolution can go. Land-based organisms have huge advantages over water-based organisms because they are able to develop a far wider range of senses. The mixture of land and sea on the Earth's surface indicates that it is not in equilibrium. If it were, then all the land would be covered by water to an equal depth. In fact, changes are constantly occurring, because of erosion, deposition, plate movements, and igneous activity. But approximate isostatic equilibrium pertains, for if the disequilibrium were too great, or there were much less water on Earth than at present, then there would be huge variations in the elevation of the land, and a far larger fraction of it would be uninhabitable and climatically extreme.

The Earth is quite different from bodies like the Moon or Mars, because almost all its surface has a very similar net gravitational force acting upon it. This is partly because so much of the surface is covered by water, and partly because very little lies more than a few hundred metres above sea-level. On planets where there are no oceans, one sees enormous variations in surface topography. The Earth's oceans and moist atmosphere both play a role in reducing the modest topographic variations by the cycle of erosion by rain, wind, and rivers that continually moves material from high to low ground. This ongoing process tends to level the surface, but is periodically overcome by mountain-building activity as a result of plate movements. The maximum height that the mountains can achieve is determined by the strength of the intermolecular forces, but the thickness and depth of the continental and oceanic crusts below them appear to be controlled by the need to maintain a global equilibrium. How this occurs, and what the limits are, is still not fully understood.

Equally crucial for the habitability of the Earth has been the evolution of its atmosphere. For half its lifetime it had a reducing or neutral chemical composition that could dissolve ferrous materials; and for the other half it has had an oxidizing composition that could transport large quantities of nonferrous metals. Combine these requirements with the need for large land-masses, so that those metals remain in an accessible form close to the surface for billions of years, and one begins to see that technologically exploitable planets are not going to be common. Moreover, when one comes to consider the existence of radioactive materials, we have been the beneficiaries of another vagary of the geological process that incorporated such materials into the Earth. Naturally occurring uranium is almost all in the form of the isotope uranium-238. (Isotopes are forms of the same element in which the atomic nucleus contains the same number of protons but a different number of neutrons.) This form of uranium will not sustain chain reactions. If you wish to construct a bomb, or a useful nuclear chain reaction, then it is necessary to extract from the uranium-238 the traces of another form of uranium, uranium-235, which can sustain an ongoing chain reaction. In naturally occurring uranium, however, no more than 0.3 per cent is in the form of uranium-235; in order to achieve a chain reaction, at least 20 per cent uranium-235 is required. (So-called 'weapon grade' or 'enriched' uranium has 90 per cent uranium-235.) The low relative abundance of the 235 isotope of uranium explains why uranium deposits and mines do not undergo spontaneous nuclear reactions culminating in huge explosions.* An abundance of uranium-235 in a usable, but safe, distribution clearly relies upon a sequence

* There is an interesting example of a natural nuclear reactor arising at a mine site in Oklo, in the African state of Gabon. In 1976, a uranium mine was discovered, which contained quantities of two isotopes of the rare element samarium. In naturally occurring samarium, the ratio of these two isotopes is usually about 9:10, but in the sample taken from the Oklo mine the ratio had been reduced to only 1:50. Conditions within the Earth at the site of the mine had conspired, over billions of years, to produce a 'natural nuclear reactor', which steadily burnt one isotope into the other. The reactor first went critical two billion years ago. In fact, the occurrence of the conversion relies on a very delicate balance between the strengths of the forces of Nature. The reactor's

1 *Fowling in the Marshes*, from the tomb of Nebamun *c.* 1450 BC.

2 Masaccio's *The Rendering of the Tribute Money*, painted in 1424–6, in the Brancacci Chapel, Church of Santa Maria del Carmine, Florence.

3 Piero della Francesca's *Flagellation*, painted in 1455, in the Ducal Palace, Urbino.

4 Georges Seurat's canvases are covered by thousands of dots of equal size. These appear to the eye as a uniform variation of colour. This 'pointillism' is illustrated in his famous work *Sunday Afternoon on the Island of La Grande Jatte* of 1884–6, in the Art Institute of Chicago. Below is a close-up displaying the detailed multipoint construction of the image.

5 Jan van Eyck's *The Arnolfini Wedding*, painted in 1434. The entire scene is mirrored in the small circular convex mirror hanging on the back wall which is shown in the detail.

6 The Helix Nebula, 400 light years from Earth. A dying star has shed a shell of hot material, which expands outwards, cooling and thinning out. Eventually, it will disperse and mingle with the ambient interstellar gas and dust.

7 Earthrise from the Moon, photographed by the crew of *Apollo 11*.

8 A typical African savannah landscape displaying the advantageous possibility of seeing without being seen.

9 Artwork displaying the hazard motif: detail from *Sailing Vessels in a Heavy Swell* by Francois Etienne Musin. The title of this work deserves an award for understatement.

10 Sunset.

11 A computer-generated fractal landscape which captures the statistical features of real landscapes which have developed by self-similar processes. Notice, however, the lack of prospect and refuge symbols (image by Richard Voss).

12 Colour for camouflage; casque head chameleon, Kenya.

13 Colour for attraction: the peacock's fan.

14 Red berries on green foliage.

15 Colour for warning: poison arrow frog, Venezuela.

16 Red-eyed tree frog.

17 Computer-generated flowers created by Przemyslaw Prusinkiewicz.

18 A Christianized seventeenth-century extension of the ancient constellations. The Northern Hemisphere and its Sky, from Andreas Cellarius, *Atlas Coelestis seu Harmonica Macrocosmica*, Amsterdam, 1660.

19 The Southern Sky, from Andreas Cellarius, *Atlas Coelestis seu Harmonica Macrocosmica*, Amsterdam, 1660.

20 Fire Beast. A striking digital image produced by Ryoichiro Debuchi.

21 The extraordinary bower of the orange-crested gardener bird (*Amblyornis subalaris*).

(a)

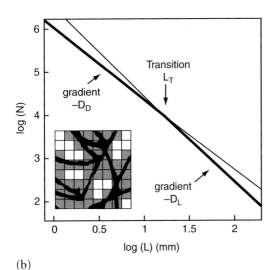

(b)

22 (a) *Blue Poles: Number 11, 1952*, enamel and aluminium paint on canvas, 210 × 487 cm, painted by Jackson Pollock in 1952 (The National Gallery of Australia, Canberra). (b) After covering a scan of *Blue Poles 11, 1952* with a computer-generated mesh Richard Taylor *et al.* counted the number of squares, N, in the aluminium layer that contain part of the pattern as a function of the size of the square L. These filled squares are shaded in the sample shown inset. The data display a characteristic of all Pollock's work with a fit to the 1000 data points given by two straight lines – one for the large-scale data, the other for the small-scale data. The straight-line character is indicative of fractal behaviour in each of these regions with a typical transition scale, L_T, of a few centimetres. These two regimes are generated by the use of two different techniques (drip and random sweeps) to create the small and large-scale patterns.

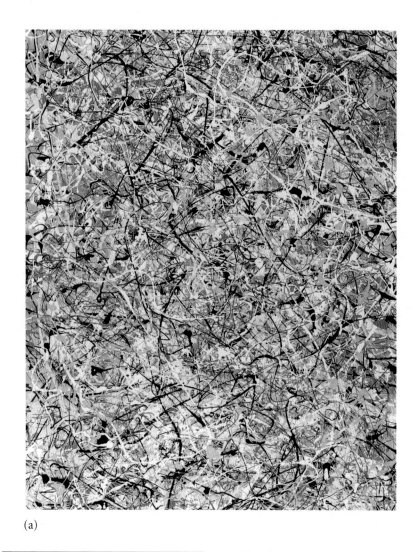

(a)

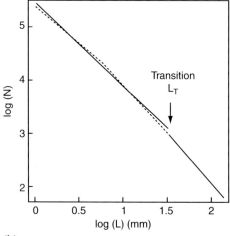

(b)

23 (a) A drip painting of unknown origin, enamel on canvas, 70 × 112 cm. (b) Analysis of the pattern complexity of the black paint layer in this unattributed painting performed by Richard Taylor *el al.* using the same method as for 22(b). Unlike for a true Pollock, the data points are not well fitted by straight lines on either small or large scales as would be the case for fractal behaviour over those distances. Two solid lines have been drawn to show the best straight-line fits to the small and large-scale behaviour so as to force a fit and create a transition length of L_T=3 cm.

of unpredictable accidents in the composition and geological evolution of a planet. One might speculate further. Imagine that the Earth was subjected to a small shower of meteors rich in diamonds, or precious metals like gold. The world economy could be thrown into turmoil; with gold now as common as iron, the gold reserves of the major industrial nations would suddenly be on the market as scrap.

The abundance of radioactive elements in the Earth's interior plays an important role in its history. They act as a source of internal heat that must be lost from the planet's surface. The rate of this heat loss determines how much of the Earth's core remains solid. As we saw in the last chapter, a small sphere has relatively more surface for its volume than does a large one. Hence, planets like Mercury and Mars, which are much smaller than the Earth, have far less internal heat build-up, and hence far less subterranean magma and vulcanism. The internal heating of the Earth plays a leading role in maintaining the plasticity of the mantle. This creates opportunities for magma to be generated and rise through the crust. If the Earth were smaller, then it would be easier to conduct away the heat from its internal radioactivity, more of the core would be solid, and volcanoes would be rarer. This decrease in the frequency of volcanic eruptions would, however, be more than compensated by the far greater impact of any that did occur. A smaller Earth would have a weaker gravitational pull at its surface, allowing volcanic dust and ash to be ejected far higher into the atmosphere. The effects on climate would be considerable; sunlight would be screened, and acids would be produced in the upper atmosphere by the condensation of sulphurous volcanic gases.

Pebble in the sky: the Moon above

It's no use telling me it's a dead rock in the sky! I *know* it's not.

D. H. Lawrence

---

products reveal that this special balance must have existed two billion years ago, just as we know it does today. Physicists were consequently able to place very strong restrictions on the possibility that the strong, weak, and electromagnetic forces of Nature could have changed their strengths very slowly over billions of years, rather than remaining constant. In the last few years the question of whether the constants of physics are all true constants has become of major interest to physicists and astronomers. Their constancy can be checked even farther back in time than the Oklo reactor by comparing the absorption patterns of light from distant quasars when it encounters clouds of dust en route from the quasar to our telescopes. These observations enable us to look back nearly 10 billion years into the past to test whether the constants that govern the interactions between matter and light have remained constant over that period of time. At present there is evidence for a very slow increase of about six parts in a million over 10 billion years from these observations, but it will take a few more years for this to be confirmed or refuted using different observations. For the fuller story, see my book *The Constants of Nature*.

The most impressive sight in the night sky is the waxing and waning of the Moon. The Moon is much larger, relative to the Earth, than is any other satellite in the solar system when compared to its major planet. Jupiter and Saturn are 317 and 95 times as massive as the Earth, respectively, but their largest moons are not much bigger than ours. The great size of the Moon has impressed itself upon our thinking about the world. From the 'lunatic asylum' to the 'man in the Moon', we see its psychological influence. But its direct physical influence upon us has been even greater. The Moon is quite close—at a distance only 60 times greater than the radius of the Earth—and its relatively large size means that the Earth and the Moon behave rather like a double planet.

All around us lunar influences have been imprinted upon our bodies by the pressures of time. The twelvefold subdivision of the year that we call the 'month' is really a 'moonth': a period close to the period of 27.32 days that the Moon takes to revolve around the Earth, in relation to the unchanging distant stars (Figure 4.3). During this period, which is called the Moon's sidereal period, the Earth will also have moved in its orbit round the Sun, and the Moon will have to move a further distance (about 27 degrees) to complete its cycle of phases relative to the Sun. In fact, allowing for this, the whole monthly cycle of lunar phases takes 29.53 days.

The presence of the Moon exerts a pull upon the Earth that is stronger on the side of the Earth that is closest to the Moon. This creates a tidal variation in the heights of the oceans, which vary monthly with the Moon's motion round the Earth. There are tantalizing hints that this variation has imprinted itself upon the behaviour patterns of living things in diverse ways. For creatures that live in shallow waters, or are amphibious, the variation of the tides provides an important variation to which adaptation will be beneficial. Women display a 28-day oestrogen production cycle, which lies close to the monthly lunar period. We call it the 'menstrual' cycle—derived from *menses*, or month. Many other mammals exhibit menstrual cycles, with associated variations in body temperature, and the time of ovulation has been found to vary between 25 and 35 days in primates. There seems to be no straightforward explanation for these correlations between the Moon's phases and menstrual cycles. Why should human fertility mirror the cycle of the Moon's changing phases? It has been suggested that it might be a vestigial remnant of an earlier stage of our evolution when our ancestors lived in the sea and were reliant in some way upon the tidal cycle. Another proposal is that these cycles are light adaptations from the period when humans were primitive hunter-gatherers. In such circumstances, daylight is a scarce commodity; the full Moon must be exploited to the full. The dark period when the Moon had waned might naturally find itself given over to mating activity, and adaptation would then occur to a body-cycle with a chemical periodicity which mirrored the lunar variation. But how such a variation could be

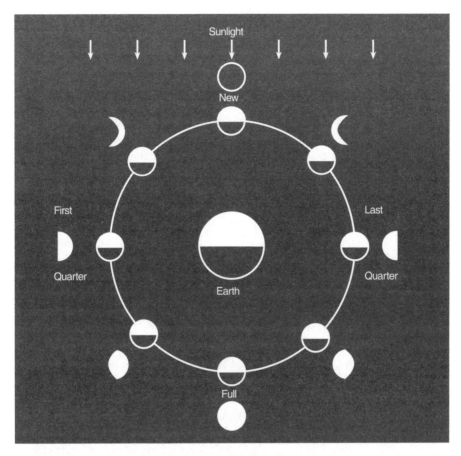

4.3 The Moon shines only in the reflected light of the Sun and so its appearance in the sky is determined by its position relative to the Sun. Half of the Moon is always illuminated by the Sun, but the part we see illuminated from the Earth varies. This picture shows what we see in the sky as the Moon passes through its various phases.

robust enough to be preserved so universally until the present, throughout so many species, remains a mystery.

Mankind's awareness of the stars, and of the periodic changes in the appearances of the Sun and the Moon, was already well developed at the dawn of recorded history. Long before written records of any sort were kept, there was an appreciation of systematic changes in the heavens. The most striking sight must have been the monthly changes in the shape of the Moon. One of the earliest human artefacts giving evidence of human counting may have been an attempt to record the lunar cycle. About thirty years ago a bone handle, which had originally been attached to a quartz engraving-tool, was found at Ishango beside

Lake Edward on the borders of modern Zaïre. It was made around 9000 BC, carved by a member of a society that lived by hunting and fishing along the shores of the lake until they were eventually wiped out by a volcanic eruption. The petrified bone handle is roughly cylindrical, and displays three rows of tally marks, as shown in Figure 4.4. The way in which the marks are grouped has fuelled considerable speculation. The top two rows both sum to 60. The third row sums to 48. There are traces of doubling, with adjacent groupings of 10 and 5, 8 and 4, and 6 and 3 marks. Moreover, the first row exhibits the sequence 9, 19, 21, 11; that is 10–1, 20–1, 20 + 1, and 10 + 1. One speculation is that the 60s represent two lunar months of days, and the marks were a calendar. The row that totals 48 is anomalous, but there have been claims that microscopic analysis reveals further markings on this section of the bone, although it is just as likely that the line is incomplete: indeed, it is to be expected that if the owner was

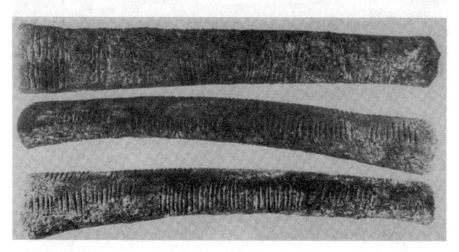

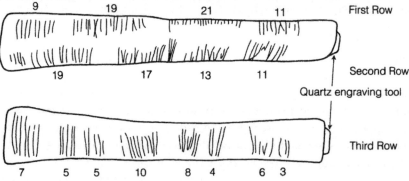

4.4 Photographs of the sides of the Mesolithic bone tool found at Ishango. It has three rows of engraved marks. The marks are indicated on the diagram beneath.

killed, or the bone was lost, that it would fall short of a whole multiple of months. We know that an accurate method of charting seasonal changes was probably important for the Ishango people, because the seasonal changes in their region forced them to leave the lakeside and migrate to the mountains when the rains came and the water level rose.

A far more ancient artefact of this sort is provided by a 30 000-year-old fragment of bone found in the early years of the 20th century at Blanchard, in the Dordogne region of France. It contains a sequence of 69 engravings on one side, arranged along a curved line that snakes back and forth five times, as shown in Figure 4.5. When examined microscopically, the marks were found to fall into groupings, and to have been made by 24 different types of tool stroke, perhaps even with different tools. This seems a laborious way to make a decorative pattern, and it seems more likely that these marks constitute some form of notation. Moreover, the crescent shapes of the markings are reminiscent of the phases of the Moon. The archaeologist Alexander Marshack believes this is what the marks are telling us, so long as we read them in the correct order, beginning with the two marks at the centre which marked the day of the last visible crescent and the disappearance of the new Moon. As one traces the marks around the curve, the full Moon is reached at the first group of four similar marks; subsequently, the full and new Moons are marked by groups of four dots, and the whole pattern is interpreted as a record of days in terms of the appearance of the Moon over a period of two and a quarter months.

In the last chapter, we saw how particular sensitivities to the natural environment would have been advantageous to an early hominid species living in tropical savannah habitats, half a million years ago. We could ask whether a response to any aspects of the heavens would offer them some advantage, to which adaptation might then occur. In this primitive world, the night was full of danger—the only time when hominids were unable to use their keen eyesight and strategic planning to outwit stronger and faster animals that had a superior sense of smell. It is easy to see why we tend to be afraid of the dark. Misfortunes would be most likely to occur at night, and so the occasions when they occurred would most naturally be associated with the shape of the Moon. The Moon and the stars would be seen when groups were gathered together around fires, talking of their hunting adventures and planning for the next day. In such circumstances, alert to the appearance of patterns in the dark, there is a tendency for the lights in the sky to become linked to the telling of tales, to acts of heroism, to exciting places, and imagined events over the horizon.

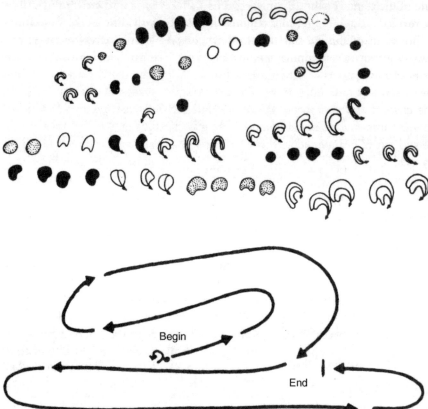

4.5 The 30 000-year-old engraved bone plaque found at Blanchard in the Dordogne, together with a chart of the pattern of marks, resembling lunar phases, and Alexander Marshack's proposed guide to the scanning order. It begins with the two marks near the centre to mark the day of the last visible lunar crescent and the first day of the invisible new Moon. Moving up the right and subsequently down the left, the full Moon is reached with the first group of four strokes at the second turn. As the line passes back to the right, the four black dots at the third bend coincide with the next new Moon. With the fourth turn, at the lower left, the count reaches another full Moon, and the final mark in a five-mark group lies near the fifth, and last, bend.

Darkness at noon: eclipses

If the stars should appear one night in a thousand years, how would men
believe and adore, and preserve for many generations the remembrance
of the city of God.

Ralph W. Emerson

Surrounded by the nocturnal glare of artificial light that bathes our cities, we see
little of the stars. For the ancients, especially those living below clear skies, or in
the rarified air of mountainous regions, things were very different. The spectacle
of thousands of shining stars would have been the most impressive thing they
saw in their lives. No wonder that myths and creation stories grew up in which
the heavenly patterns of light played a leading role. Eventually, amazement must
have given way to familiarity, only to be rejuvenated occasionally by unpredictable changes in the heavens. At the beginning of this century, the philosopher
George Santayana delivered a famous series of lectures in the United States on
the subjects of beauty and aesthetics. He picked out the appearance of the night
sky as an exemplar of what is attractive to the human mind: a level of intricacy
delicately poised between unfathomable complexity and uninteresting simplicity. The hint of pattern challenges the mind to ponder and seek it out. So,
what if we were to see the night sky for the first tine? Emerson's words, at the
head of this section, imagine the spiritual consequences of such an astral
awakening. They inspired the young Isaac Asimov to pen his famous short story
Nightfall about the final days of the civilization on the planet Saro. This world
enjoyed the light of six suns. At least one of them was always high in the sky.
Natural darkness was unknown; and so, therefore, were the stars. The inhabitants had evolved in a world of light with no psychological conditioning for
darkness, and a strong susceptibility towards claustrophobia when deprived of
light. Their astronomers were convinced of the smallness of the Universe.
Unable to see beyond their own sixfold solar system, they contented themselves
with showing how well its complicated motions could be understood using the
same law of gravitation that worked so well on the surface of Saro. These ration-
alists shared their world with romantic Cultists, who perpetuated an 'old knowledge' of a world of light beyond the sky, and a coming day of darkness when the
world would end. Many discounted the Cultists as irrationalists, but others saw
their beliefs as a confused tradition arising from a past appearance of darkness
and celestial lights in the sky, long ago. Social tensions mount when the astronomers predict that there must be an unseen dark moon in their solar system that
will become visible only when it comes close to eclipsing one of their suns. Its
presence is required to explain the intricate motions of the suns. A few astronomers realize that the moon will eclipse the second sun of the system, at a time

when it is the only sun in the sky. The eclipse will be total. News of this expectation leaks out. Civil unrest mounts as the Cultists stir up eschatological fever; the eclipse begins to eat away at the disc of the solitary sun; it ends in totality. Darkness blots out the sky and tens of thousands of brilliant stars appear, shrouding the planet in a canopy of twinkling starlight. For Saro is not a denizen of the sparsely populated stellar suburbs of a galaxy like the Milky Way; it lies deep in the dense heart of a star cluster. Panic and civil unrest break out. There the story ends; the reader is left to ponder the revolution in outlook that is about to occur.

Looking for parallels in history, we might compare the impact of the first appearance of that star-studded darkness on the fictional world of Saro with early human responses to a complete eclipse of the Sun by the Moon. Ancient eclipses are famous for their influence upon human affairs. The total eclipse that occurred on 28 May in 585 BC was so dramatic and unexpected that it ended the five-year-old war between the Lydians and the Medes. Their records tell us that in the midst of battle 'the day was turned into night'; their fighting stopped immediately and a peace treaty was signed—endorsed by marriages between their royal families. In stark contrast, the eclipse of the Moon on 27 August in 413 BC brought about a rather different end to the Peloponnesian War, between the Athenians and the Syracusians. The Athenian soldiers were so terrified by the eclipse that they became reluctant to leave Syracuse, as planned. Interpreting the eclipse as a bad omen, their commander delayed the departure for a month. This delay delivered all their forces into the hands of the Syracusians: they were totally defeated, and their procrastinating commander was put to death.

Many centuries later, Christopher Columbus exploited his astronomical knowledge of an eclipse of the Moon by the Earth to enlist the help of the Jamaicans after his damaged ships were stranded on their island in 1503. At first, he traded trinkets to the natives in return for food; eventually, they refused to give him any more, and his men faced the prospect of starvation. His response was to arrange a conference with the natives on the night of 29 February 1504—the time when an eclipse of the Moon would begin. Columbus announced that his God was displeased by their lack of assistance, and He was going to remove the Moon as a sign of His deep displeasure. As the Earth's shadow began to fall across the Moon's face, the natives quickly agreed to provide him with anything he wanted, so long as he brought back the Moon. Columbus informed them that he would need to go and persuade his God to restore the lesser light to the heavens. Retiring with his hour-glass for the appropriate period, he returned, in the nick of time, to announce the Almighty's pardon for their sins and the restoration of the Moon to the sky. Soon afterwards, the eclipse ended. Columbus had no further problems on Jamaica; he and his men were subsequently rescued, and returned in triumph to Spain.

Eclipses are remarkable things. Their existence has influenced cultures the world over for thousands of years. They have found their way into art, into theology, folklore, and astrology. Their notability ensured that ancient historians invariably recorded them, and often interpreted them as omens of great significance. This makes them useful as a method of dating written accounts very precisely. For instance, in the biblical book of the prophet Amos, he writes (chapter 8, verse 9) of Nineveh, 'And it shall come to pass in that day, saith the Lord God, that I will cause the sun to go down at noon, and I will darken the earth in the clear day'. The 'day' in question was 15 June 763 BC, and is also recorded in the Assyrian state chronicles after being observed at Nineveh.

Eclipses occur because of an accident of Nature (Figure 4.6). The true diameter of the Sun is about 400 times greater than that of the Moon; its distance from the Earth is also about 400 times greater than that of the Moon. These gross disparities conspire to make the apparent sizes of the Sun and the Moon in the sky the same. As a result, the passage of the Moon in front of the Sun

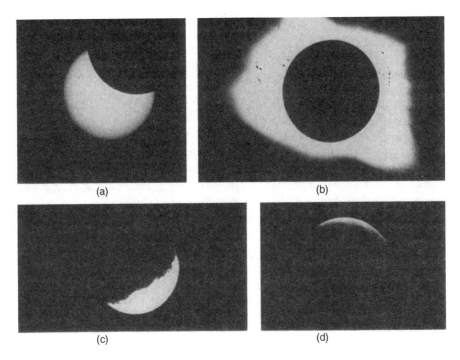

(a)   (b)

(c)   (d)

4.6 Solar and lunar eclipses: (a) a partial eclipse of the Sun in which the new Moon has cut off part of the visible solar surface; (b) a total eclipse of the Sun; (c) a partial eclipse of the Moon in which the Earth's shadow obliterates about two-thirds of the visible surface of the Moon; (d) a near total eclipse of the Moon by the Earth's shadow.

can cover the face of the Sun completely, to produce a total eclipse of the Sun. By way of contrast, if we examine the other planets in the solar system, we find that their moons will appear much larger than the Sun in their skies. On the average, our Moon appears to be just a little smaller than the Sun when viewed from the Earth. But the difference is small enough to be overcome by the variations in the distance between the Earth and the Moon, so that there are also periods when the face of the Moon is slightly larger than that of the Sun. The situation is finely balanced: if the distance to the Moon were increased by a mere 8 per cent (about 29 000 kilometres) then total eclipses of the Sun would never be seen from the Earth.* Now, we have already explained that the distance between the Earth and the Moon is gradually increasing by a few centimetres every year. Five hundred million years from now, the Moon will be so far away that total eclipses of the Sun will be no more. We live at a propitious time for eclipse-watchers. But, as we shall see in a moment, the fortuitous accidents of time and space that permit us to see total eclipses have more far-reaching consequences.

Eclipses were always bad news for the ancients. Even when advanced cultures understood why they were occurring, they continued to endow them with a meaning that was bound up with human events. The word 'eclipse' derives from the Greek *ekleipsis* meaning an 'omission' or an 'abandonment', and in many other cultures there are remnants of an old image of the Sun being consumed by a wild beast during an eclipse. In Chinese, 'to eclipse' is *shih*, 'to eat', with the Sun being devoured—traditionally by a dragon. But for modern astronomers eclipses are not bad news. The coincidence that the Sun and the Moon have the same apparent sizes in the sky, despite the vast difference in their true sizes, has been of the deepest significance for the progress of our understanding of the Universe. Before we see why, recall the many speculations about the inevitability of long-lived extraterrestrial civilizations becoming scientifically advanced. Let us suppose that the deep, unifying 'Theory of Everything' that modern physicists are searching for truly exists. Let us even assume that mathematics is a universal language of Nature that is appropriate for expressing that Theory of Everything. Thus, any full understanding of Nature, any deep exploitation of Nature's potential, must come about through an understanding of those mathematical laws that underpin the workings of the Universe. This is, of course, a comforting philosophy for those listening for, or sending, extraterrestrial signals. The search for signals from extraterrestrials is based upon a belief in the universality of mathematics and the laws of Nature. This does not mean that we expect

---

* The only other place in the solar system where a complete eclipse would be seen is from Prometheus, an irregular satellite of Saturn. But from the surface of Saturn, so far from the Sun, the eclipse of the Sun would extend over a tiny area of sky and be very brief in duration.

extraterrestrials to use the same alphabets, or number systems, that we do. But it is believed that, none the less, they must describe, by some means, the same basic logical connections as our own systems, and so they will be able to translate our description into their own—just as we are able to converse about numbers with people from other cultures using translation. This is why the messages we send—so hopefully—into space use wavelengths of light that have special significance for physicists. The significance of these wavelengths should be appreciated by anyone whose knowledge of matter and radiation enables them to send or receive radio signals. It is an interesting question how reasonable are all the assumptions behind such far-reaching expectations. But let us grant them all for now, because we are more interested in another unnoticed assumption: that advanced civilizations, of similar intelligence to ourselves, will be able to deduce the laws of Nature as easily as ourselves. We tend to think of ourselves as likely to be about average in the celestial IQ rankings, admittedly pulled up a lot by the occasional Einstein, who departs so far from the average (Figure 4.7). We also tend to think of 'advanced' as an across-the-board accolade: if they know a lot about anything, they will know a lot about everything.

Any more-advanced civilization stands a good chance of being older and smarter than we are today. Physicists like Ed Witten have made the assumption referred to above, and have argued that, given enough time, others would have to converge upon a Theory of Everything, if it exists. But perhaps not. Our own progress in science has been expedited at many a turn by some remarkable coincidences in our situation in the Universe. The equality of the apparent sizes of the Sun and the Moon is a remarkable case in point.

One of our clearest glimpses of part of a Theory of Everything is provided by Einstein's remarkable theory of gravitation: the general theory of relativity. This was first announced in 1915, 228 years after Newton's original law of gravitation was published. Newton's classic law works beautifully in all practical circumstances on the Earth, because gravity is comparatively weak. But when very strong gravitational forces are encountered, they can bend the paths of light rays by significant amounts, and Newton's theory fails to account for what is seen. In these situations, Einstein's theory succeeds with breathtaking accuracy. But the distinctive differences between the predictions of Einstein's theory and those of Newton's simplified theory are very small: even on the scale of the solar system, they amount to no more than one part in one hundred thousand, and they are observable only in unusual circumstances.

Einstein's theory predicts that, when the light from a distant star grazes past the surface of the Sun, its path will be bent as if it were feeling the pull of the Sun's gravity. The extent of this 'light-bending' is very small, and the only circumstances in which we can hope to see it are those created by a total eclipse of the Sun. During an eclipse, astronomers can determine which distant stars

4.7 A Herblock cartoon from the *Washington Post* published on 18 April 1955, the day of Einstein's death.

can be seen and which are eclipsed. Since their positions in the sky at any moment can be predicted very precisely, one can determine by how much the distant starlight has been bent by the Sun simply by noting the positions of the stars that would have been eclipsed if light travelled in straight lines (Figure 4.8). Without the coincidence that creates total eclipses of the Sun for us, this prediction of Einstein's general theory of relativity could not have been tested. Einstein made these predictions about the bending of starlight in 1916, during the First World War. Fortunately, there was an eclipse in 1919, soon after the war ended, and it occurred in front of the best star-field for testing the light-bending predictions.

The other great success of Einstein's theory for our understanding of the solar system, which served to confirm the essential truth of the theory to astronomers of the time, also hinges upon a quirk of the solar system. When the planets orbit

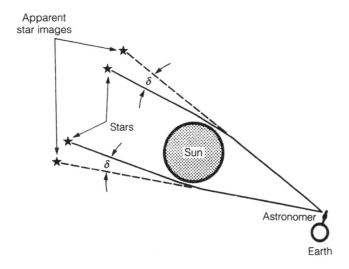

4.8 The gravitational deflection of light rays from distant stars as they pass close to the Sun. The deflection is the angle between the observed position of a star in the sky during the total eclipse (when the light is deflected by the Sun's gravity) and its position when the Sun is elsewhere in the sky (and so the Sun's gravitational effect is negligible). For the Sun, this deflection (the angle δ) is about 0.000 486 degrees.

the Sun they do not move in perfectly elliptical orbits because of the perturbations they experience from the others. The ovals just fail to join up, and the next orbit traces a similar oval, which is just slightly displaced from the previous orbit. Eventually, the path of the planet would trace out a rosette shape, as shown in Figure 4.9. We say that the oval orbit 'precesses'. The amount of precession

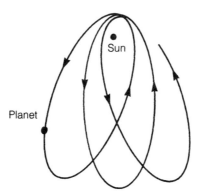

4.9 A precessing orbit. The orbit of the planet is approximately an ellipse which rotates its orientation.

can be measured by the angle between successive extremities of the orbit. Some contributions to this precession have been known since Newton's day. The largest comes from the tugs that the orbiting planet receives from the gravitational pulls of all the bodies in the solar system other than the Sun. But, by the end of the nineteenth century, an embarrassing problem had established itself. After all the known perturbations to its orbital path had been accounted for, the orbit of the planet Mercury displayed a mysterious unexplained residual precession. It amounted to a precession of just 43 seconds of arc* per century.

Einstein's theory of general relativity predicted tiny (one part in 100 000) corrections to Newton's classical predictions concerning the orbits of planets around the Sun. In effect, very close to the Sun, where the pull of the Sun's gravity is strongest, there are tiny deviations from Newton's famous law of gravitation, which had predicted that the strength of the Sun's gravitational force should fall as the square of the distance from its centre. Einstein's theory predicted that the correction to Newton's law should produce a precession of Mercury's orbit amounting to 43 seconds of arc per century—exactly what was required to explain the long-standing discrepancy. Now this precession occurs for all planetary orbits, but its magnitude depends upon the distance of the planet from the Sun. The farther the planet is from the Sun, the smaller is the amount of precession that the Sun's gravity creates. For all the planets in our solar system other than Mercury (the closest to the Sun), the precession is too small to be observed. If our solar system had not contained a planet as close to the Sun as Mercury, it would not have altered the course of events that led to the evolution of intelligent life on Earth, but we would have been robbed of a unique opportunity of checking the truth of Einstein's theory of gravitation.

The twofold coincidence of Mercury's closeness to the Sun and the visibility of eclipses from the Earth, occasioned by the similarity in the apparent sizes of the Moon and the Sun, have had the most profound consequences for the development of human understanding. Because of these two accidents, we were able to test the theory of gravitation to great precision, and use it with confidence farther afield in the Universe. Without these coincidences, we would have been left for half a century with Einstein's beautiful theory as a monument to human ingenuity, without any way of discovering whether it was true or false. Thus we see how accidental aspects of an extraterrestrial civilization might have subtle and far-reaching consequences for their intellectual advancement. If you live on a large lone planet, circling a star like the Sun then, in order for conditions to be cool enough to support life, it must be so far away from the star that its orbital precession is too small to allow you to uncover a better theory of

---

* There are 360 degrees in a circle, 60 minutes of arc in a degree, and 60 seconds of arc in a minute of arc. A second of arc is denoted by ".

gravitation than Newton's. Without a very specially situated moon, of just the right size, you will see no total eclipses, and you will be unable to learn of light-bending by your star's gravity. And, indeed, without other planets you have only a cyclopean view of the whole business of planetary formation.

The lesson of this little example is simple. It should not be assumed that extraterrestrials, no matter how cerebrally advanced, will inevitably discover all the approximations to the laws of physics that will ultimately converge upon a Theory of Everything. Many of those discoveries require the presence of environmental configurations in which the differences between simple and better approximations to the true laws of Nature are manifest. All that is needed is a cloud-covered planet for a civilization to develop a wonderful understanding of meteorology, without an inkling of astronomy. An absence of lodestone, or a rate of planetary rotation that is too slow to create a noticeable magnetic field, means that the development of an understanding of magnetism would be greatly retarded. A quirk of geology could mean that radioactive elements were absent, or buried inaccessibly deep underground: the result would be an impediment to an understanding of weak and strong nuclear forces. Of course, it is easy to think of clever ways in which we could overcome such restrictions to our knowledge if they were suddenly imposed upon us here and now.* This does not really matter. We could never have taken the difficult first steps along the road that reached our present state of knowledge without the unique possibilities that the quirks of our position in the Universe have provided. Scientific knowledge among civilizations of fairly similar levels of maturity will quite probably be very uneven. It will reflect the vagaries of their local environment and the problems that needed to be overcome in order to survive more comfortably for long periods before any scientific investigation took place. The frequency of wars will play a significant role in the speed of technological advance. The level of understanding each civilization possesses about the scale of the Universe, and the nature of its contents, will be the most susceptible to truncation by poor visibility. We must remember that while there are evolutionary reasons for living things to further their understanding of their local environment—survival prospects are, for example, enhanced by understanding motion, electricity, immunology, and radioactivity—no such advantage appears to be offered by knowing that the Universe is expanding, or that black holes exist. One day, perhaps, we may find

---

* The best way of checking the light-bending predictions now is to look at the bending of radio waves emitted by very distant sources (near the edge of the visible Universe), whose position can be measured with great accuracy. This method does not require an eclipse, but the distant sources of radiation are quasars, and so we find ourselves taking advantage of another lucky coincidence: that the Sun passes in front of two quasars so that their radiation goes close enough to the Sun to be deflected by a measurable amount. This is done by measuring the change in the angle between the two quasars as they pass behind the Sun.

one. If we do, it may not be an advantage that is simple and direct. I suspect that some other practical advantage will be possible only as a by-product of this more esoteric knowledge.

▨▨ Hamlet's mill: the wandering Pole Star

> And [Jacob] dreamed, and behold a ladder set up on the earth, and the top of it reached to heaven: and behold the angels of God ascending and descending on it.
>
> Genesis 28: 12–13

The Earth is not alone, spinning on its axis in the depths of space. The Moon and the Sun conspire to create one further peculiar effect upon the motion of the Earth. The Earth's rotation causes it to develop a midriff bulge around the equatorial regions, where the outward rotational forces are greatest. Since the Earth's axis of rotation is tilted with respect to the plane of its orbit around the Sun, the equatorial bulge of the Earth is not located in the plane of its orbit either. As a result, a force is exerted upon the Earth by the gravitational field of the Sun, which tends to move the Earth's axis so that its bulge lies in the plane of its orbit (Figure 4.10).

In addition, the equatorial plane of the Earth is not aligned with the plane of the Moon's orbit; and, because it is closer to the Earth, the Moon exerts an even stronger torque upon the spinning Earth than the Sun. The effects of these forces

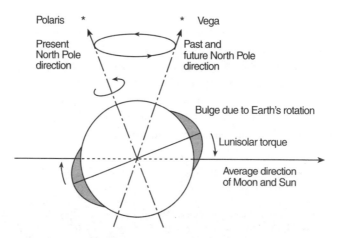

4.10 The precession of the Earth's axis of rotation. The pull of the Sun and the Moon on the Earth's equatorial bulge (shown exaggerated here) causes the rotation axis of the Earth to precess slowly around the Pole of the ecliptic about every 26 000 years with respect to the distant stars.

upon the Earth are similar to those we see when we push a spinning top. Instead of simply changing the direction of its axis of rotation, we cause the direction in which the axis is pointing to rotate, or precess, in a circle. The gravitational pull of the Moon and the Sun on the equatorial bulge of the Earth has a similar effect, and so the direction of the Earth's North Pole slowly changes. It takes about 26 000 years for the Pole to complete its circle of precession and return to point in the same direction. Tradition holds that this phenomenon was first discovered by the Greek astronomer Hipparchus in 125 BC. It is believed that he compared the celestial positions of stars as observed by himself with the positions as recorded by others two centuries earlier, and discovered that they had shifted systematically. (However, later in this chapter, we shall suggest that he may have been alerted to this in another way.)

One of the consequences of the Earth's precession is to change the direction in which the North Pole points. At present, we are rather fortunate. The position of the Pole Star (Polaris) marked by its companions, the two 'pointers', is a very close approximation to the true position of the exact celestial Pole. By contrast, there is no conveniently placed star in the southern sky to mark the South Pole's direction in the sky. 'Polaris' is the Latin for 'of the Pole', and derives from the Greek word *polos* meaning a pivot or an axis, although this was not used by astronomers until the Renaissance. We are rather fortunate, because Polaris is one of the brighter stars in the sky—indeed, the brightest that is within about half a degree (the size of the full Moon) on the sky over the entire 26 000-year path of the Pole's precession. For most of this path, there was no close candidate to use as a Pole Star at all; but Polaris is at present only about 44 minutes of arc away from the true North Pole. The Greeks and the Romans had no Pole Star. Shakespeare, writing in 1599, has Julius Caesar say that he is 'constant as the Northern star', but this is a complete anachronism. Hipparchus tells us, in about 125 BC, that 'at the pole there is no star at all'. Figure 4.11 shows the path of the North Pole's direction against the stars, during the past and in the future. A thousand years from now, Vega will take its turn as the star that is closest to the direction of the Pole; but navigators will find it a poor substitute for Polaris, because it will be several degrees away from the true Pole.

The Poles have a deep significance for many who watch the sky. They create an axis around which the entire sky appears to revolve. In Figure 4.12, this is vividly displayed in a long-exposure photograph showing star trails tracing circular paths centred on the axis of the Pole. One can see how well Polaris (conveniently near the top of the tree in the photograph) marks the central point around which all other stars move. Not surprisingly, for the ancients and for the peoples of many traditional cultures, this celestial rotation and the direction round which it turned, held deep and magical significance. The Pole was the one stable fixed thing in the heavens amidst a sea of movement that threatened to displace

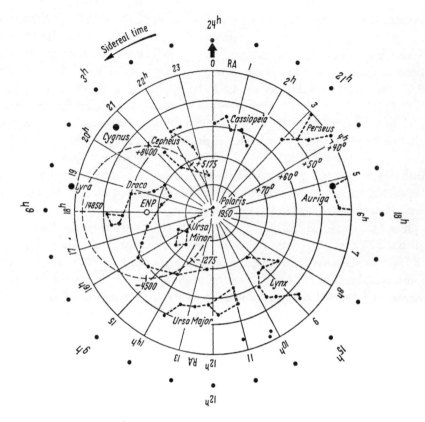

4.11 The path of the North Celestial Pole as seen from a latitude of 50 degrees north (e.g. that of Prague or Frankfurt). The position of the Celestial North Pole is traced (dashed) for various dates before and after AD 1950. At present it is very close to Polaris, the Pole Star.

it and bring the sky down in chaos. The early Egyptians saw it as a celestial avenue, leading to eternal life. In many Scandinavian and North Eurasian cultures, the Pole Star is named the 'Nail Star' to emphasize its fixed position— 'nailed' to the sky. For the Imperial Chinese, the Pole pointed to the throne of the sovereign of the cosmos, around whom the stars were arrayed.

By virtue of its profound status at the centre of the sky, the nearest star to the Pole became the well-spring of legends of many sorts. Their ubiquity inspired two historians, Hertha von Dechend and Giorgio de Santillana, to ascribe a vast body of ancient mythology and legend to cataclysmic prognostications about the great sky axis. They entitled their book *Hamlet's Mill*, in recognition of the many ancient traditions that likened the circling of the stars round the celestial Pole to the grinding motion of a millstone. One finds this motif in many of the legends

4.12 A long time exposure directed at the North Celestial Pole records star trails left by the northern stars as they follow their circumpolar paths. The North Celestial Pole is the only point on the sky that does not move; it is very close to Polaris, our Pole Star, which is conveniently sited above the apex of the tree in this photograph by Michael McDermott.

of Siberia and Scandinavia. In the first century before Christ, we find Greek astronomers referring to the Pole as a place where 'the heavens turn around in the way a millstone does'. Armed with this mythological focus upon a magic mill in the sky in a variety of cultures, together with its symbolization of stability and wealth, von Dechend and Santillana set about interpreting innumerable myths and fables the world over as coded identifications of the significance of the sky axis. They try to argue that there is a homogeneity to many human myths and beliefs held by disjoint cultures—and thus to the cultural leanings that they induce—that is brought about by the shared significance they attribute to the sky axis. This is a theme that is potentially wider than their study. Extraterrestrial civilizations will almost certainly be constrained to live in solar systems sharing many of the features of our own: a similar stable star, about the same distance away so as to render conditions temperate enough to support life, and rotating about an axis pointing towards two preferred ('north' and 'south') directions on the sky. There might well arise myths, speculations, and stories not dissimilar in emphasis (although differing, of course, in their particulars) to those we find on Earth. While the authors of *Hamlet's Mill* undoubtedly get carried away in their quest for an astronomical underpinning to every human myth and legend under the sun, and their mountain of historical information in parts represents little more than a miasma of hopeful associations, their book contains a core of truth. The lesson it teaches us is that the shared human experience of the heavens is one that imprinted itself upon our imaginations in pre-scientific times. Myths are often attempts to join the heavens and the Earth. Impressive celestial sights, whether they be of the Moon, of the Sun, or of the sky axis around which the world turns, are shared human experiences for many. It is no accident that they form the basis of so much human fantasy and religious yearning.

The myths of the sky axis are all to be found in cultures that live in northerly latitudes. There is a profound reason for this: a reason that has had far wider consequences for humanity's growing awareness of the celestial scenery. The night sky in the tropics is quite different from that of more temperate latitudes. For a long time, explorers and anthropologists were puzzled by the differences between the astronomical systems developed by sophisticated tropical cultures, developed over the last two thousand years, and those found in Europe and North America. They failed to appreciate the different character of the sky at low latitudes. As we have already discussed, when seen far from the Equator the stars appear to rotate around the celestial Pole, giving it the appearance of the centre of things. The greater the latitude, the higher the celestial Pole will be in the sky. From northern latitudes all the heavenly motions appear to be centred on the Pole; fewer stars can be seen, but many of them are always visible; they can therefore be used for timekeeping and wayfinding. Tropical skies do not look like this. An observer there finds that the movements of the stars mirror the rotation of the Earth. At the Equator, every star can be glimpsed, although the celestial Poles are lost at the horizon. Stars rise, reach their zenith and then fall in the sky and set. As a star rises, its direction remains relatively constant and provides an excellent navigational 'fix' for a long period of time. There is very little horizontal motion, and the sky appears very symmetrical. For this reason, one finds that many Oceanic cultures developed linear constellations that tracked the rising paths of stars. By contrast, as one moves to northerly latitudes, stellar motions become a mixture of vertical and increasing horizontal components, and the sky appears more asymmetrical. The appearance of the night sky is thus, in many ways, simpler for the tropical observer. He appears to be at the centre of things, beneath a celestial canopy of overarching movements that he can use to fix directions of travel (Figure 4.13).

For our Pleistocene spectators in Africa, there would have been no apparent Polar axis; the stars would have passed overhead, making them feel at the centre of the world. However, whereas there is an adaptive advantage to be gained by a sensitivity to the periods of the Moon—so that moonlit nights can be exploited for hunting, and vigilance can be increased on dark moonless nights when the danger from surprise attack is at its greatest—no such advantage was offered to

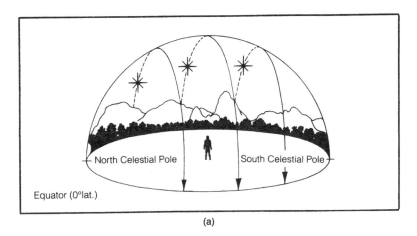

(a)

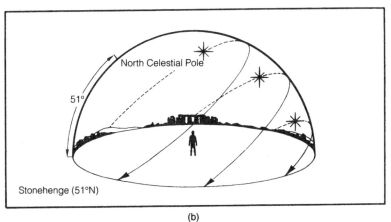

(b)

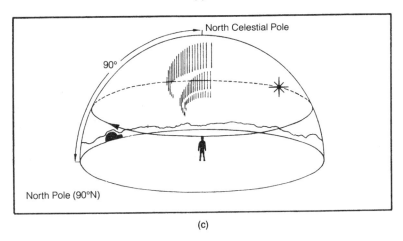

(c)

4.13 The variation in the appearance of the night sky with the latitude of the observer because of the change in location of the celestial Pole around which the stars appear to rotate: (a) at the Equator; (b) at temperate latitudes; (c) at the North Pole.

primitive star-gazers. On the darkest nights, the light from thousands of stars might offer some comfort and security. But a fascination with fire is a far more useful propensity to engender. Firelight offers protection from the beasts, whereas moonlight and starlight help you to be seen as well as to see. Land-going savannah-dwellers did not need to navigate at night. Only long-distance, nocturnal travellers and mariners needed to study the stars. But perhaps an interest in the stars is an inevitable by-product of a fascination with the Moon. Since a response to the phases of the Moon does offer adaptive advantage, sensitivity to the light of the Moon and of the stars would naturally be engendered in survivors.

Paper moon: controlling chaotic planets

Perhaps that was the necessary condition for planetary life: Your Sun must fit your Moon.

Martin Amis, *London Fields*

We have seen how the tilt of the Earth's axis—the obliquity of the ecliptic—is the source of our seasonal variations. Even small changes in this obliquity can have potentially catastrophic consequences for our climate. For a long time, it has been suspected that perturbations to the 26 000-year period of the Earth's precession, caused by the Moon or the planets, could create small changes in the angle of obliquity that, if they amounted to only about one degree, would be sufficient to explain the occurrence of ice ages. This theory was first proposed, sixty years ago, by a Yugoslavian, Milutin Milankovitch, while he was a prisoner of the Austro-Hungarian Empire during the First World War. (The Hungarian Academy of Sciences still allowed him to pursue his studies in Budapest.) He argued that past changes in the Earth's rotation and obliquity would alter the amount of solar energy reaching different parts of its surface at high latitudes, and produce variations in temperature and glaciation that could be correlated with geological evidence of past ice ages. More recently, the past behaviour of the Earth's obliquity has been illuminated by new studies, by Jacques Laskar and his colleagues in Paris, which reveal the importance of the Moon's presence for the Earth's habitability.

Over very long periods of time, the rate at which the Earth's Polar axis precesses (currently about 50″ per year), its obliquity, and the shape of the Earth's orbit around the Sun, all change slightly in response to the increase in distance of the Earth from the Moon, and to the gravitational influences of the other planets. At present, the effect is very small: the obliquity is changing at a rate of only about 47″ every century. But if this change were to be extrapolated backwards for even half a million years, the change in the Earth's obliquity would be enormous—more than 65 degrees—and the climatic changes utterly devastating: the tropics would cease to exist. Fortunately for us, extrapolating the current rate of change of the obliquity backwards like this is not a reliable indication of what happens to it over hundreds of thousands of years. Its behaviour is far more complicated. In order to determine how the obliquity evolves, we must consider other aspects of the Earth's motion to which it is tied. The most important is the rate of precession, which is determined by the length of the day, because it is a measure of the rotation rate of the planet. What makes the long-term evolution of the obliquity potentially eventful is the phenomenon of 'resonance'. We are familiar with it in many mundane situations. If we push a child on a swing, then there is a particular frequency for pushing that creates an especially large swinging response. This is a resonance; it occurs in any situation where the frequency at which an outside disturbance is applied matches the system's natural frequency of oscillation. The consequences can sometimes be devastating, as they were for the infamous Tacoma Narrows Bridge in Oregon, which collapsed after the resonant amplification of torsional oscillations of the bridge that occurred in high winds. When other planets perturb the Earth with a frequency equal to its precession rate, resonances occur, and can create a change in its obliquity over only tens of thousands of years. Since the distance between the Earth and the Moon is steadily increasing, at a rate of about 3 centimetres per year (roughly, as fast as your fingernails grow), many of these resonant interactions could have occurred in the past, when the Moon was closer and the Earth was spinning faster.

Detailed computer simulations of the evolution of the rotation, precession, and obliquity of all the planets in the solar system have revealed a remarkable situation. The obliquity of a planet can evolve chaotically over long periods of time, changing by large amounts, in response to small perturbations, because of its extreme sensitivity to the combined effects of resonant perturbations, changing rate of rotation, and the distortion of the shape of the planet that accompanies changes in its rotation rate. Before considering the Earth, it is interesting to see the results for Mars. Mars is a simpler object for study because it has no moons large enough to play a significant role in the evolution of its spin and obliquity; its rotation is likely to be primordial, left over from the conditions that accompanied its formation. It precesses at 8.26″ per year, which is close to the frequency of some of its natural vibrations. As a result, its obliquity is expected to have varied chaotically all over the range from 0 to 60 degrees (see Figure 4.14). Its present obliquity of 24 degrees could therefore have arisen from a starting value anywhere within that wide range. The chaotic sensitivity of its precession means that we cannot reconstruct its past history before 100 million years ago, and hence determine its initial obliquity: the uncertainties in its present motion eventually overwhelm any attempt at further extrapolation into its past. This is

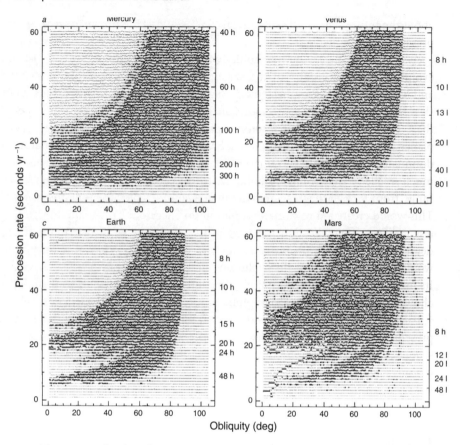

4.14 The ranges of a planet's obliquity and rate of spin precession that lead to chaotic or regular evolution of its obliquity over time. The spin precession is given in units of seconds of arc per year on the left-hand vertical axis, with the corresponding rotation period shown on the right-hand vertical axis. The cases of Mercury, Venus, Earth, and Mars are shown. The dark regions are those displaying chaotic variations; the lighter ones signify regular evolution of the obliquity. In a chaotic zone, the obliquity can vary anywhere along a horizontal line within the zone. Typically, the entire width of a chaotic zone will be explored erratically in a few million years. The present situation of the Earth, with the Moon present nearby, is represented by a point with precession rate of 55 arc seconds per year and an obliquity of about 23 degrees. This lies comfortably within the regular zone. Current values for each planet's obliquity (tilt) and rotation period can be found in Table 4.1, on p. 141.

the classic situation for a 'chaotic' physical system. Although we may be in possession of an exact law that predicts the future of the system from its past, any uncertainty in the specification of its past state will amplify so quickly that the exact law becomes less and less useful; ultimately, it gives no information at all about the future state of the system. Likewise, the past cannot be found from the present.

A similar chaotic evolution holds for Mercury and Venus. In contrast, the evolution of the obliquity of the large outer planets (Uranus, Jupiter, Saturn, and Neptune) is much more stable, because their precession rates are far smaller (less than 5″ per year) and strong resonant effects hardly ever occur. In between these extremes of chaos and stability, which distinguish the inner and outer planets, sits the unique case of the Earth. The evolution of its obliquity is dominated by the presence of the Moon. If the Moon did not exist, or was much smaller, then the Earth's obliquity would evolve chaotically over the entire range from 0 to 85 degrees, remaining above 50 degrees for millions of years. This would create a dire climatic situation on Earth: the Poles would receive far less radiation than the Equator. Given that past variations of just one or two degrees have been sufficient to trigger ice ages, variations of this magnitude would be catastrophic for the evolution of life. Fortunately, the Moon does exist. Its presence acts as powerful stabilizing influence, and its gravitational influence allows the Earth's obliquity to do nothing more dramatic than oscillate by about 1.3 degrees about its mean position of 23.3 degrees* (see Figure 4.15). The present period of obliquity decrease is just a downturn in the oscillatory sequence. One day it will reverse. We cannot, however, conclude that the Earth's obliquity has always wobbled around its present value, because the Moon may not always have been present. There could have been a period of chaotic evolution of the obliquity prior to the capture of the Moon by the Earth's gravitational field or creation from the results of an impact with the proto-Earth 4600 million years ago.† After that capture, its changing obliquity would have been shepherded by the Moon towards a future of stable oscillations around a value of 23.3 degrees. A possible thermal history for the two cases is shown in Figure 4.16.

These discoveries display the crucial importance of a lunar presence over very long time-scales. The moderate climatic variations of Earth are linked to the levels of tilt and rotation that the Earth possesses. Over long periods of time, the

---

* Some studies indicate that chaotic changes in obliquity might be stabilized, even in the absence of the Moon, if the Earth were rotating fast enough, with days shorter than 8 hours. This could occur because a high level of rotation increases the equatorial bulge of the Earth; lunar tides have a similar effect.

† Because of the difficulties of affecting a capture and the compositional similarities of the Moon and the Earth with respect to some isotopes, the 'impact theory' is currently favoured by planetary scientists. This theory proposes that the Moon arose from an impact between the proto-Earth and another body. A near grazing blow would allow the core of the incident body to accrete on to the Earth's core, while its mantle would mix with the Earth's in a vapourized form. Some of this material would rain down on the Earth's surface while the rest condenses gravitationally to form the Moon. This would explain the smallness of the Moon's core.

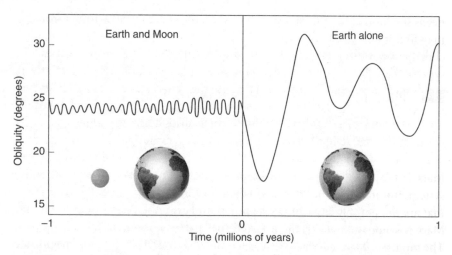

4.15 The expected time variation in the obliquity of the Earth with the Moon present (left) and absent (right). The presence of the Moon leads to stable small variations (±1.3 degrees around an average value of 23.3 degrees). If the Moon were absent, large and irregular variations would occur. The right-hand example was computed by removing the Moon from the left-hand calculation of the history at time 0.

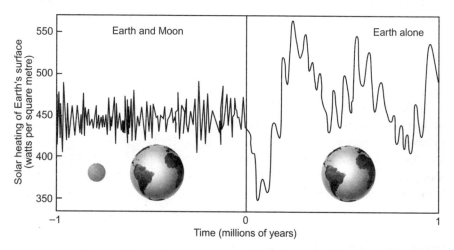

4.16 The expected time variation in the solar heating of the Earth's surface at a latitude of 65 degrees north with the Moon present (left) and absent (right).

precession of the Earth's polar axis is driven by its rotation rate and, in conjunction with its obliquity, responds occasionally to the other bodies in the solar system. Those responses would be erratic, changing dramatically every 100 000 years or less, were it not for the pacifying presence of the Moon. Stable climate needs the Moon; other worlds on which the evolution of complex life has begun may find it extinguished, or eternally hamstrung, by the need to adapt to huge climatic variations unless their planet, too, has a moon for a dancing partner.

Mars in your eyes: they came from outer space

The aim of science is not to open the door to infinite wisdom, but to set a limit to infinite error.

Bertolt Brecht

Mars is the fourth planet from the Sun and the seventh largest in the solar system. It is named in honour of Ares the mythical Greek god of war for reasons that are not entirely clear, but perhaps because of its red colour. In our cultures Mars is synonymous with the extraterrestrial. The word 'Martian' is common. The month of March derives from Mars, and whole sections of the confectionary business are devoted to selling Mars chocolate bars. This is curious. Great planets like Saturn, Jupiter, and Neptune have a much lower terrestrial profile. Do their inhabitants just employ the wrong marketing company, or could there be something about Mars that makes it so much more fascinating for the average Earthling? How did it become the epitome of an alien world?

Mars is easily visible with the naked eye in the night sky. Its brightness varies a lot, as does its distance from the Earth. Every 26 months Mars approaches its closest to Earth and we can send a space probe there with a minimum of fuel. This is why in 2004 we seemed to be seeing spacecraft from Europe and America queuing up to land and orbit the red planet. It was one of those times of closest approach.

Although it is rather smaller than the Earth, Mars has about the same amount of surface land. It has two tiny moons, Phobos and Deimos, that look like misshapen potatoes: Phobos is a mere 22 kilometres and Deimos a trifling 12 kilometres in diameter. Both are simply asteroids that got too close to Mars and found themselves captured by its gravity.

Our fascination with Mars has been fed by the tantalizing patterns visible on its surface. In the autumn of 1877, when Mars was also close to Earth, the great Italian planetary astronomer Giovanni Schiaparelli (the uncle of the famous fashion designer Elsa Schiaparelli) at the Brera Observatory in Milan thought he saw channels (*canali*) on the surface of Mars. When his reports were translated into English, *canali* became *canals*, suggesting that they had been artificially constructed by local Martian residents for purposes of irrigation or transport. The dark and light areas on the planet surface he named after terrestrial seas, capes, and peninsulas, coining exotic and euphonious names like the Herculis Columnae (Columns of Hercules), Aurorae Sinus (Bay of the Dawn), and Solis Lacus (Lake of the Sun). By these leaps of imagination Schiaparelli had recreated

Mars in the image of an ancient Earth, pregnant with myth and meaning. It was never the same again.

Intrigued by Schiaparelli's drawings and detailed observational reports, the American astronomer Percival Lowell added weight to this misconception. In 1894, he claimed that the intricate grid of surface markings was the result of work by intelligent beings who were inhabiting the planet even now! Lowell developed his views in three books: *Mars* (1895), *Mars and Its Canals* (1906), and *Mars As the Abode of Life* (1908), and by then Mars had become the most fascinating place in the solar system.

This speculative groundwork laid the foundations for the work of great science fiction writers like H. G. Wells and Olaf Stapleton and a host of their successors who continue unabated today. So enthusiastically did the American public seem to take up the idea of intelligent Martians that on Sunday 30 October, the day before Halloween, 1938 the young Orson Welles was able to create panic amongst millions of Americans who tuned in to a portion of his radio adaptation of Wells' *War of the Worlds*. They became quickly convinced that they were hearing reports of a real Martian invasion of America! A huge flaming object had landed in New Jersey. News broadcasts came on within the story, read by actors describing the Martians as they emerged from their spaceships:

They look like tentacles to me. There, I can see the thing's body. It's large as a bear and it glistens like wet leather. But that face. It . . . it's indescribable. I can hardly force myself to keep looking at it. The eyes are black and gleam like a serpent. The mouth is V-shaped with saliva dripping from its rimless lips that seem to quiver and pulsate . . . . The thing is raising up. The crowd falls back. They've seen enough. This is the most extraordinary experience. I can't find words. I'm pulling this microphone with me as I talk. I'll have to stop the description until I've taken a new position. Hold on, will you please, I'll be back in a minute.

Eventually, the real newsreels had to appeal for calm and explain the mass panic.

Today, it is we who are 'invading' Mars. Detailed observations long ago revealed that Lowell's canals were just tricks of the human eye, which has evolved to detect patterns, joining up neighbouring points to make lines whenever it can. Yet, the meandering channels are real. In early 2004 we had evidence from the *Mars Express* space probe that there is frozen water at the South Pole of Mars and that flowing water probably once eroded great channels in its surface. Perhaps deep below the surface the pressure of packice is great enough to sustain liquid water even now.

For astronomers, Mars teaches us about the wonderful properties of Earth. Mars has no plate tectonics: its terrain is simple. Also, unlike the Earth, Mars has no magnetic field. This left the Martian atmosphere at the mercy of the fast-moving electrically charged particles that are blown towards it from the Sun. Gradually, they blew away the Martian atmosphere, leaving almost nothing behind. Earth's atmosphere would have suffered the same fate had it not been for our magnetic field. It deflects the incoming wind of solar particles around the atmosphere so we hang on to it.

Mars had a far more extreme climatic history than Earth. The reason is again remarkable, as we have just seen. Both the Earth and Mars rotate with an axis of rotation inclined at about 23–24 degrees from the vertical to the plane of their orbits around the Sun. But without the benefit of the stabilizing effect of a large Moon, and unable to hang on to its atmosphere, Mars has been subjected to this chaotic climatic history, as is witnessed by the huge variations in ice and temperature around its surface. Without the Moon, complex life on Earth would perhaps only have been able to exist, like that on Mars, in the minds of other beings and on the pages of their science fiction books.

In the future our exploration of the solar system will focus with ever greater emphasis upon the surface of Mars. And as it does so, the aura of Mars will be embellished with new images of a world that was once living but which dies, perhaps seeding life on Earth and playing one last part in the creation of worlds like Earth that can know about themselves.

The man who was Thursday: the origins of the week

> I must have a prodigious quantity of mind; it takes me as much as a week, sometimes, to make it up.
>
> Mark Twain, *The Innocents Abroad*

The day, the month, and the year are periods of time replete with celestial meaning. If, by some oversight, we lost track of these cycles, all would not be lost. Our timekeeping could soon be reinstated because it is anchored to periodicities in the heavens which, while not perfectly constant, are constant enough over long intervals of time for all practical purposes. Terrestrial, lunar, and solar timescales first impressed themselves upon Earth-dwellers in ways that are independent of culture; subsequently they provoked elaboration and celebration in accord with the multitude of cultural responses to time. Since they reflect real periodicities in the terrestrial environment, they create a variation to which adaptations are possible by degrees. Our bodies bear the hallmarks of daily and monthly changes; our world displays the annual pattern of the Earth's motion around our local star, and ebbs and flows in an endless dance with the Moon. But not all our divisions of time are impressed upon us so directly from outside. There are aspects of our experience that have been structured indirectly by our interpretation of the heavenly motions, rather than by those motions directly. The most pervasive is our habit of parcelling groups of days into convenient small periods that we call weeks. Today, a division of seven days is universal, and in many languages the word for week is simply that for 'seven days'.* Where did this ubiquitous division come from?

The lunar month does not divide into equal multiples of a whole number of days; yet there is clearly something astronomical about the days of the week— Sun-day and Mo(o)n-day are undeniably celestial—although there is no exact seven-day cycle of change on view in the heavens. The week seems to be a largely cultural creation. In some non-Western cultures, 'weeks' originally had different lengths, and in the past some Western totalitarian regimes have attempted, unsuccessfully, to redefine the length of the week. The story of how our week arose is a curious one because it displays an unexpected merger of two opposing influences. The first is an attempt to resist celestial influences upon human affairs, while the other is the embracing of astrological influences. We shall find that the days of the week have much to tell us about the historical processes that culminated in their present names.

The earliest division of time with no link to the phases of the Moon was that of the ancient Egyptians. As devout Sun-worshippers, they had reason to exclude any lunar influence from certain aspects of their social structure. They divided the year into twelve 30-day months, each of which was subdivided into three 10-day weeks, leaving five special days to be fitted in throughout the year. This division of the year into 36 weeks seems to have been primarily astrological in significance; each week was associated with a particular constellation of stars whose rising coincided with the first day of the week.

If we look for the source of the Western tradition of the seven-day week there are two, possibly entwined, threads of history. On the one hand, there is the Jewish tradition of the seven-day cycle of creation, ending with the sabbath day of rest; on the other, we find, in Babylonia and Chaldea, the rise of astrological relations between the seven ancient planets. Both these sources lie in the same geographical region and might flow from a common, more primitive source. Some have argued that the Jewish sabbath tradition, and the creation stories of Genesis, arose during the period of their exile in Babylon, following the destruction of Jerusalem in 586 BC. The Jewish adoption of the seven-day cycle was linked to particular considerations of national identity and exclusive theology. Whereas other nations in the region had strong astrological practices, and made a habit of worshipping the Sun and the Moon as divinities, this practice never seems to have arisen among the Jews. For them, the adoption of a time cycle that was not tied to the Sun or the Moon was one way of

* For example, in French (*semaine*), Spanish (*semana*), Greek (*hebdomas*), or Hebrew (*shavu'a*).

shunning the worship of the Sun and the Moon, and reinforcing their belief in the status of those bodies as created things. This was an important part of the evolution of their religious thinking towards the recognition of God as wholly other and unrepresentable by created materials. If the weekly cycle had been based upon some other period (say a quarter of the Moon's cycle) then the veneration of the sabbath would have found itself associated with a natural celestial cycle.*

Although this was the final manifestation of Jewish sabbath observance, there are other biblical traces of an earlier link between the sabbath and the phases of the Moon. There are four Old Testament passages that suggest this vestigial connection. In the first (2 Kings 4: 23), the Shunammite woman's husband asks her when she is planning to visit Elisha, 'Wherefore wilt thou go to him to-day? it is neither new moon nor sabbath.' Since her journey requires the use of an ass, one interpretation of his question is that it is usually available only on the sabbath, when it is not required for farm work; so, she is expected to make her journey on that day. Alternatively, since she is visiting the prophet to seek his help in healing her son, perhaps the new Moon or the sabbath are propitious times to seek the prophet's intercession. In Isaiah (1: 13), the 'new moon and sabbath' are mentioned in a list of unsatisfactory religious observances. In Hosea (2: 11), a warning is issued to Israel by Yahweh: 'I will cause all her mirth to cease, her feasts, her new moons, and her sabbaths, and all her solemn assemblies.' And in Amos (8: 5), the prophet denounces traders who decry the restrictions that religious observance imposes upon their trading hours, saying 'When will the new moon be gone, that we may sell corn? and the sabbath, that we may set forth wheat?'

These references have led to claims that the sabbath may have originally been

---

* It is intriguing that in the fifth tablet of the Babylonian creation narrative *Enuma elish*, featuring the solar deity Marduk as a parent of the created world, and often compared with the Hebrew creation story, there is a hint of the week being linked to a quarter of a month. It reads

> The Moon he caused to shine, the night (to him) entrusting.
> He appointed him a creature of the night to signify the days:
> Monthly, without cease, form designs with a crown.
> At the month's very start, rising over the land,
> Thou shalt have luminous horns to signify six days.
> On the seventh day be thou a half-crown.
> At full moon stand in opposition in mid-month.
> When the sun overtakes thee at the base of heaven,
> Diminish thy crown and retrogress the night.

On the seventh day a 'half-crown' is variously translated as with 'half a tiara' or 'halve thy disk', implying that the seven-day cycle was linked to the visual appearance of the Moon as it passed from a tiara-like crescent shape to its half-moon shape. Unfortunately, this hint that the month was divided into four seven-day periods is not made any more explicit than this in the rest of the text.

the day of the full, or the new, Moon. Later, the Jews held a feast on the new Moon, but not on the occasion of the full Moon. This is consistent with some early full-Moon celebration having been absorbed and superseded by the more frequent pattern of sabbath observance. There is no doubt that there was a lunar cycle of social observances. The practice is prescribed in Numbers 28: burnt offerings are to be made at the start of every month. They greatly exceed the size of those required on the sabbath. But this does not help us decide whether the monthly celebrations preceded the sabbath ones. The Book of Genesis makes no mention of either observance.

The Jewish tradition marked the seven-day cycle of the week by the observance of the sabbath as a day of rest and religious worship. In time, this aspect of the week has come to dominate the structure of Western societies. Its most interesting testament is the wedge that it drives between human affairs and the structure of Nature. When life is organized around a schedule created by human symbolism, it is freed from the strictures of Nature, and a certain spirit of independence is engendered. For the Hebrews, this ancient practice was established to reflect their beliefs about the pattern of creation. Yahweh acted creatively for six days, and then rested on the seventh. The word 'sabbath' is derived from *shabath*, meaning 'to cease from work', whereas the Hebrew word for week (*shavu'a*) is allied to that for seven (*sheva*). The sabbath was dedicated to God and became the fulcrum about which all social and religious activities turned. Its precise origin has proved impossible to pinpoint, but some scholars have drawn attention to ancient Babylonian records of things that were forbidden, even to the king, on every seventh day; there is a similar Babylonian word *shabbatum*, or *shapattum*, meaning the 'day of the rest of the heart', with a related meaning to the Hebrew. It is not clear, however, whether these taboos applied only during special months; nor do they appear to have been very prohibitive. Inspection of huge numbers of dated Babylonian commercial documents reveals that there was no reduction in the number of transactions carried out on these seventh days when compared with others. If there was a Babylonian seven-day cycle, it had a different orientation from that of the Hebrews. The similar Hebrew and Babylonian words for sabbath may point to a common origin for both of them. That origin would most probably have been a marking (by celebration or abstinence) of the new or full Moons, with intermediate quarters gradually producing lesser observances. The Hebrews took over this pattern, injecting it with a special significance of their own to emphasize their national solidarity and exclusivity in the face of possible dilution by cultural influence and intermarriage. None the less, a residual connection with lunar festivals remained and re-emerged at times when their religious observance relapsed. Yet their observances must have been quite distinct from those of the Babylonians at the time of the exile, because they staked their national and religious distinctiveness from the Babylonians upon the practice of sabbath observance.* Its prominence in the Decalogue, second only to their obligations to Yahweh, displays the importance that was attached to it.

Babylonian astrological practices have, in fact, proved to be just as pervasive as the institution of the Jewish sabbath. Our description of the days of the week derives from the intricacies of those beliefs and practices. This is betrayed by the obvious connections between the names for the days of the week in many European languages, and those of the seven ancient 'planets'—Saturn, the Sun, the Moon, Mars, Mercury, Jupiter, and Venus—displayed in Table 4.2. In the ancient world, the 'planets' (or 'wanderers') in the sky included the Sun and the Moon, together with the five other members of the solar system visible to naked-eye observers. In languages influenced by Latin one can see many of the Roman names for the days mirroring the names of the ancient planets. In others, like English and German, the translation process has adopted the corresponding Norse gods or goddesses as replacements for the Roman gods corresponding to the planets. Thus, *Thursday* (Thor's day) in English and *Donnerstag* (*Donars-tag*, Donar's day) in German have replaced Jupiter, the Roman god of the sky, by Thor or Donar, the Norse god of thunder, who is also sometimes known as Thunar.

In all these languages we see the direct correspondence between the days of the week and the seven ancient planets at the heart of astrological interpretation, rather than with the days of the Hebrew creation story, which culminated in the institution of the sabbath. The Babylonian and Chaldean astrological system ascribed to each of the celestial bodies that 'wandered' with respect to the stars a god who controlled aspects of human affairs. An explicit association of planets with days can be found in early Babylonian horoscopes dating from about 410 BC. The subsequent arrival at the present system, and the way in which the names of the planets are ordered into our sequence of named days, is clearer, but curiously elaborate. By the second century BC, a conventional ordering of the seven planetary bodies had been set in place. It was dictated by the hierarchy of their speeds in the heavens. The fastest movers had the shortest orbital periods when viewed from the Earth (remember that it was assumed that all these bodies, even the Sun, orbited around the Earth). This gives the following

---

* Many societies developed a pattern of rest days which were linked to taboos, often timed to coincide with seasonal changes, and with the phases of the Moon. The Hawaiians had strict taboo-days when no fires could be lit, silence was observed, no canoes were launched, no bathing took place, and people went outdoors only for religious observances. Because of the lunar connection, the system used is not dissimilar to that of the sabbath with four taboo periods in each month. The Hawaiians singled out the period between the 3rd and 6th nights, the full Moon (including the 14th and 15th nights), the 24th and 25th nights, and the 27th and 28th nights. It is not uncommon to find abstinence being practised at the time of the new and full Moons, and consequently for these days to be dedicated to some deity.

Table 4.2 Words for days of the week that have an astronomical root across a variety of European languages. See also Figure 4.17.

language	Saturn	Sun	Moon	Mars	Mercury	Jupiter	Venus
Latin	dies Saturni	dies Solis	dies Lunae	dies Martis	dies Mercurii	dies Jovis	dies Veneris
Cornish	de Sadarn	de Sil	de Lûn	de Merh	de Marhar	dê Jeu	de Gwenar
Breton	Disadorn	Disul	Dilun	Dimeurz	Dimerc'her	Diriaou	Digwener
Welsh	dydd Sadwrn	dydd Sul	dydd Llun	dydd Mawrth	dydd Mercher	dydd Iau	dydd Gwener
Gaelic	Di-sathuirne		Di-luain	di Màirt			
Catalan			Dilluns	Dimarts	Dimecres	Dijous	Divendres
French			Lundi	Mardi	Mercredi	Jeudi	Vendredi
Italian			Lunedi	Martedi	Mercoledi	Giovedi	Venerdi
Spanish			Lunes	Martes	Miércoles	Jueves	Viernes
Romanian			Luni	Marţi	Miercuri	Joi	Vineri
English	Saturday	Sunday	Monday	Tuesday	Wednesday	Thursday	Friday
Swedish		Söndag	Måndag	Tisdag	Onsdag	Torsdag	Fredag
Danish		Sondag	Mandag	Tirsdag	Onsdag	Torsdag	Fredag
Norwegian		Sondag	Mandag	Tirsdag	Onsdag	Torsdag	Fredag
Icelandic		Sunnundagur	Mánudagur				
Finnish		Sunnuntai	Maanantai	Tiistai		Torstai	
Sami			Manodag	Tisdag		Tuoresdag	
Dutch	Zaterdag	Zondag	Maandag	Dinsdag	Woensdag	Donderdag	Vrijdag
German		Sonntag	Montag	Dienstag		Donnerstag	Freitag
Albanian	e Shtunë	e Dielë	e Hënë	e Martë	e Mërkurë	e Enjte	e Prëmtë

descending sequence (with their approximate periods relative to Earth in brackets):

Saturn (29 years)
Jupiter (12 years)
Mars (687 days)
Sun (365 days)
Venus (225 days)
Mercury (88 days)
Moon (27 days)

One might have expected this ordering to dictate the sequence of days. If so, the pattern in English would be Saturday, Thursday, Tuesday, Sunday, Friday, Wednesday, Monday. But the actual sequence is different. It is obtained by beginning at any day and then jumping over the names of two planets to get the next day. So, for example, starting with Saturday we skip Thursday and Tuesday to get Sunday; then skip Friday and Wednesday to get Monday; then (returning to the start) jump Saturday and Thursday to arrive at Tuesday; and so on, until all seven days have been picked out and we return to Saturday.

The contents page of a work of the historian Plutarch, dating from AD 100, lists a work by him entitled *Why are the days named after the planets reckoned in a different order from the actual order?*, but the work itself has been lost. A later discussion by the Roman historian Dio Cassios tells of an astrological practice, which probably had its origins in Alexandria. The doctrine of 'chronocracies' assigned each one of the twenty-four hours of each day to one of the seven planetary gods. The god controlling the first hour of the day also had the added distinction of being named the controlling 'regent' of that day. Each person's life was believed to be controlled, hour by hour, by the appropriate deity, or 'chronocrater', under the aegis of the regent governing that day.

These two astrological beliefs are what established our sequence of days. There were twenty-four hours in each day, and seven gods associated with the seven planets. The first hour of the first day would be assigned to Saturn, the most distant planet. Each subsequent hour is then assigned to the planets in accord with their descending orbital time periods: Saturn—Jupiter—Mars—Sun—Venus—Mercury—Moon—Saturn—Jupiter—Mars—Sun ... and so on, indefinitely. But because 24 is not exactly divisible by 7 (there is a remainder of 3), the twenty-fifth entry in the sequence, which is assigned to the first hour of the second day, is the Sun; the forty-ninth entry, which marks the first hour of the third day, is the Moon; the seventy-third entry, which marks the first hour of the fourth day, is Mars; the first hour of the fifth day is Mercury, the first of the sixth is Venus, and the first hour of the seventh day reverts to Saturn again. The sequence of planets assigned as regents to the first controlling hour of each twenty-four-hour day gives the order of days in the astrological week, which we retain to this day: Saturday—Sunday—Monday—Tuesday—Wednesday—Thursday—Friday, and so on cyclically, as shown in Table 4.3.

The early development of the Jewish and astrological weeks was quite separate after possible points of contact at their inception. But their common seven-day period ensured that they would eventually merge into a common system distinguished only by the meanings ascribed to particular days. By the first century AD there was a link between the sabbath and Saturn's day. Interestingly, the strength of the Jewish sabbath tradition is displayed by the fact that the Hebrews named the planet Saturn *Shabtai*, after the original Hebrew word for the sabbath. Thus the astrological practice of naming the days after the planets was inverted in this single case. Yet the astrological week spread far and wide from Alexandria in the second century BC. The empires of Alexander the Great and of the Romans brought together the great ancient cultures of learning around the Mediterranean and West Asia. All these cultures were linked by astrology, and readily adopted the pattern of the astrological week. This tradition was eventually taken up by both Christianity and Islam, and it spread with their converts. But astrology spread more quickly through the Roman Empire than Christianity, and its grip was so strong that, even when Christianity was adopted, there was no hope of renaming the days of the week to sever them from their pagan origins. It is interesting to note that the astrological assignment of the weekdays remains complete in languages like Welsh, English, and Dutch, which were spoken at the margins of the Roman Empire, and so were among the last to feel Christian influence during the first centuries AD. By contrast, the languages spoken nearer the heart of the Empire, where the influence of Christianity spread rapidly and more strongly, reflect the desire to express aspects of the Christian religion by replacing the astrological names of days with new ones of religious significance (Figure 4.17 and Table 4.2).

The clearest example of this is the removal of any association between our Sunday and the Sun. This day had become the first day of the week for Christian believers who, if they were also Jews, endowed it, like the sabbath (Saturday), with a special status. Its religious significance derives from its being the day on which the Resurrection occurred—hence its subsequent description in the Early Church as 'The Lord's Day'. In Latin, this translates directly into *dies Dominica*, and thence into Italian (as *Domenica*), French (as *Dimanche*), Spanish and Portuguese (as *Domingo*), and similarly in many other languages. In some languages, like Russian, the word for Sunday is just 'resurrection' (*Voskresénie*). Likewise, the influence of the Jewish sabbath can be found on other languages, displacing Saturn's day with *sabbato* in Greek, with *sabato* in Italian, and with *samedi* in French.

Table 4.3 The Babylonian sequence of planetary days. The sequence of seven 'planets' runs in decreasing order of their orbital period on the sky and therefore begins with Saturn. The 'planet' that falls on the first hour of each successive day is designated the astrological ruler of that day, and the ensuing sequence of seven daily rulers generates the order of days in the astrological week that we still use today.

1 SATURN	20 Jupiter	14 Jupiter	8 Jupiter
2 Jupiter	21 Mars	15 Mars	9 Mars
3 Mars	22 Sun	16 Sun	10 Sun
4 Sun	23 Venus	17 Venus	11 Venus
5 Venus	24 Mercury	18 Mercury	12 Mercury
6 Mercury		19 Moon	13 Moon
7 Moon	1 MOON	20 Saturn	14 Saturn
8 Saturn	2 Saturn	21 Jupiter	15 Jupiter
9 Jupiter	3 Jupiter	22 Mars	16 Mars
10 Mars	4 Mars	23 Sun	17 Sun
11 Sun	5 Sun	24 Venus	18 Venus
12 Venus	6 Venus		19 Mercury
13 Mercury	7 Mercury	1 MERCURY	20 Moon
14 Moon	8 Moon	2 Moon	21 Saturn
15 Saturn	9 Saturn	3 Saturn	22 Jupiter
16 Jupiter	10 Jupiter	4 Jupiter	23 Mars
17 Mars	11 Mars	5 Mars	24 Sun
18 Sun	12 Sun	6 Sun	
19 Venus	13 Venus	7 Venus	1 VENUS
20 Mercury	14 Mercury	8 Mercury	2 Mercury
21 Moon	15 Moon	9 Moon	3 Moon
22 Saturn	16 Saturn	10 Saturn	4 Saturn
23 Jupiter	17 Jupiter	11 Jupiter	5 Jupiter
24 Mars	18 Mars	12 Mars	6 Mars
	19 Sun	13 Sun	7 Sun
1 SUN	20 Venus	14 Venus	8 Venus
2 Venus	21 Mercury	15 Mercury	9 Mercury
3 Mercury	22 Moon	16 Moon	10 Moon
4 Moon	23 Saturn	17 Saturn	11 Saturn
5 Saturn	24 Jupiter	18 Jupiter	12 Jupiter
6 Jupiter		19 Mars	13 Mars
7 Mars	1 MARS	20 Sun	14 Sun
8 Sun	2 Sun	21 Venus	15 Venus
9 Venus	3 Venus	22 Mercury	16 Mercury
10 Mercury	4 Mercury	23 Moon	17 Moon
11 Moon	5 Moon	24 Saturn	18 Saturn
12 Saturn	6 Saturn		19 Jupiter
13 Jupiter	7 Jupiter	1 JUPITER	20 Mars
14 Mars	8 Mars	2 Mars	21 Sun
15 Sun	9 Sun	3 Sun	22 Venus
16 Venus	10 Venus	4 Venus	23 Mercury
17 Mercury	11 Mercury	5 Mercury	24 Moon
18 Moon	12 Moon	6 Moon	
19 Saturn	13 Saturn	7 Saturn	1 SATURN

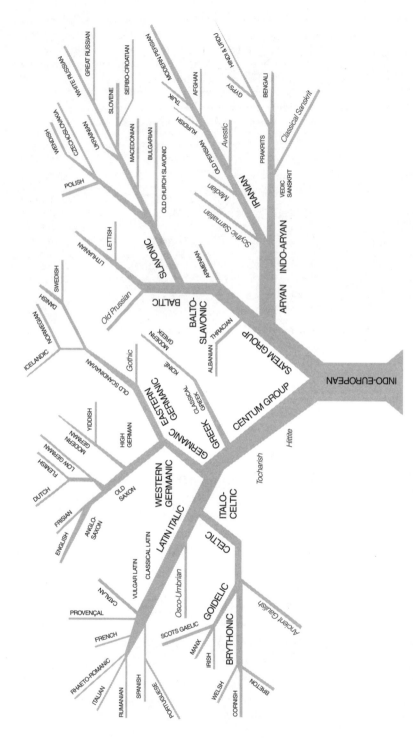

4.17 The evolutionary tree of Indo-European languages.

The dual observance of Saturday and Sunday in the Judaeo-Christian world ended officially around 360, when sabbath observance ceased. The decision of the Christian Church to maintain a separate identity by fixing its special day of rest on Sunday, to distinguish it from the Jewish sabbath, displays the importance for religious movements of special calendrical identities. The same trend is found in the foundation of Islam. Mohammed chose Friday as the holy day of the week, presumably copying the feature found in Judaism and Christianity. The spread of Islam to Africa and Asia took with it the seven-day structure of the astrological week. Thus, we still find a widespread distinction of the days Friday, Saturday, and Sunday throughout the Western world, and throughout the colonial regions of the New World, reflecting the formative influences of the Islamic, Jewish, or Christian traditions. One of the most striking (and frustrating) things for any visitor to Jerusalem is the confluence of these religions in the various quarters of the old city. Different monuments and churches open and close on different cycles, and the whole city seems to work for only four days a week. I remember being told that this was a factor that always slowed the pace of Arab-Israeli negotiations.

The most interesting feature of a holy day is its influence upon social and civil affairs. Whenever a state has been minded to eradicate religious influence, it has targeted the pattern of the week, and hence the distinguished holy day, without which the religion's community of worshippers would be disorganized and debilitated. There have been two dramatic, although ultimately unsuccessful, examples of the state waging war upon religious traditions in this way in European history. The first was the French plan, between 1792 and about 1799 to decimalize time. After the French Revolution of 1789, there was a desire for revolutionary change in other areas of life. French scientists and mathematicians had created the metric system of weights and measures, which we still use today. Others saw this as an opportunity to promote the decimalization of time, by introducing a division of each month into three ten-day cycles called *décades*.* This left the year with five special days, to be taken after the last month of summer, together with a sixth every leap year. The system is similar to that adopted in ancient Egypt, which we mentioned earlier. In order to cement the revolutionary calendar of the new French Republic, the twelve months were renamed, using names to describe a typical climatic characteristic or farming activity of the month.

The new 'Revolutionary Calendar' was introduced by official decree on 24 November 1793. Further decimalization was suggested in order to divide each day into ten decimal hours of 100 decimal minutes, each of 100 decimal seconds'

---

* Official documents began to use prosaic new names for the ten days in the decadic cycle: *primidi, duodi, tridi, quartidi, quintidi, sextidi, septedi, octidi, nonidi, décadi.*

duration. This reform was enunciated with the intention of superseding the astrological logic at the heart of the seven-day week. Moreover, it was stated that the new calendar should not resemble that used by the Roman Catholic or other apostolic churches. One of its aims seems to have been the abolition of the religious observance of Sunday. The ensuing conflict between the Catholic order of the Dominicans (named from the Latin *dies dominica*, or 'the Lord's day'), and the 'Decadists', was a result of this aim. Opposition to the observance of Sunday became draconian during the Reign of Terror, when the closing of businesses, the donning of special Sunday dress, and the opening of churches on the old seven-day Sunday cycle were all forbidden. In 1794, Robespierre attempted to institute a new state religion dedicated to the worship of the Supreme Being each *décadi*. His aim was to alter the centre of gravity of French life, and replace the influence of the Church by that of the State. However, after reaching its zenith in 1798, the whole enterprise gradually disintegrated, and it was virtually non-existent by the end of the eighteenth century.* Its failure was officially recognized by Napoleon's official reinstatement of the seven-day week, together with Sunday as the day of rest, in September 1805. The Gregorian calendar, already in use in Britain and America, and still used universally today, was readopted.

The other notable attempt to reform the week was Stalin's institution of the 'uninterrupted production week' in the Soviet Union in 1929. Here, there was a twofold purpose. One was to avoid a fallow day once a week, when all machinery would lie idle and all production cease; the other was to disrupt the pattern of family and community life to an extent that traditional religious observance would be unsustainable. Stalin set about achieving these ends by introducing a five-day cycle with four days of work followed by one day of rest. The cycle was not the same for everybody. The rest days were staggered throughout the population, so that factories and farms were constantly in production, with 80 per cent of the population working, and 20 per cent resting, on any given day. At first, each of the days of the new production week was labelled by a number, but the numbers were soon replaced by colours. Individuals began labelling their friends, family members, and acquaintances by their 'colours'. Society fragmented into five chromatic sub-societies. The 'yellows', who had their day off on the first day of the week, could socialize only with other yellows. Families were fragmented because different rest days were allocated to different members of the same family. Attempts at religious observance were thwarted by the lack of opportunity for whole families or communities to meet together on the same day.

---

* Another difficulty arose from the adoption of the day of the autumn equinox in Paris as the first day of the year. This would have led to discrepancies with other astronomical systems, because the French initiative was not taken up by other nations.

Despite close attention by the authorities, the 'uninterrupted production week' eventually degenerated into uninterrupted weak production. Workers whose duties, friends, and responsibilities were compartmentalized into a single day began to value their work very little. The absence of key workers who were needed to maintain equipment played havoc with the goal of continuous production. By 1931, the internal tensions were becoming acute and Stalin suspended the reform, blaming the irresponsibility of the workers and promising the reintroduction of the production week after a process of re-examination and re-education. But it was never reintroduced, and the whole idea was killed by his decree two years later. However, as if to emphasize the conflict with religious tradition, it was not replaced by the traditional seven-day week. Instead, it gave way to a six-day week—albeit with a single universal day of rest. This scheme continued to meet with resistance that grew in strength the farther one strayed from the centre of government. Peasant communities followed their hallowed seven-day cycle wherever possible, regardless, and eventually the State gave up, reinstating the seven-day cycle with the traditional 'day of resurrection' as the day of rest on 26 June 1940.

These battles for the seven-day week and its day of religious observance are instructive. They reveal the power of cultural tradition to order our lives. History shows that the structuring of days in a weekly cycle enables religious faiths to establish their identity by the device of hallowing particular days, or introducing a particular practice on particular days (for example, the former Roman Catholic tradition of abstaining from meat on Fridays). One should remember that there is nothing astronomically necessary about the cycle of days being sevenfold. If one steps into cultures in Africa, Asia, and the Americas that were outside the sphere of influence of the early Jewish tradition and of Mesopotamian astrology, then one finds 'weeks' of other lengths. In Africa and Central America, the weekly cycle is often framed around agricultural communities and trade. The market day is the most important day, and the weekly cycle of life revolves around it. In some parts of Africa, the word for 'week' is that for a 'market'. Another interesting feature of the length of weekly cycles in some non-Western civilizations is their link to the base of counting system used.* Distinctive examples are to be found in Central and South America, where counting systems based upon 20 (the number of fingers on two hands plus the toes on two feet) rather than our own 'decimal' system based upon 10 (the number of fingers on two hands) were widespread. Both the Mayans and the Aztecs employed base-20 counting systems and 20-day time cycles to define their weeks; the Mayans chose to divide their year into eighteen 20-day weeks and five additional, special days.

---

* A detailed account of the nature, diversity, and evolution of different counting systems can be found in my earlier book *Pi in the Sky: Counting, Thinking, and Being* (Clarendon Press, 1992).

We have dwelt upon the origins of the week because it is an all-pervasive social institution whose *raison d'être* is unknown to most people, although it dominates the pattern of our daily lives. Its source is more subtle than that of the day, or the year, or the seasons, and its role in structuring religious identity is striking; it combines traces of lunar origins, but its present form manifests the ancient influence of astrology as a way of organizing the human perception of events in the heavens. Modern astronomers find no evidence for any astrological link between the stars and human activities; none the less, the fact that such a connection was widely believed to exist in the past was reason enough to frame the pattern of human activities and determine the names of the days of the week throughout Western cultures. Again, almost without noticing, we have found the heavens imprinting themselves upon our ways, if indirectly; this time, through the desire of our forebears to imbue their motions with meaning and to link the progress of time on Earth with the will of the gods.

Long day's journey into night: the origin of the constellations

We are all in the gutter, but some of us are looking at the stars.

Oscar Wilde, *Lady Windermere's Fan*

There is one piece of astronomy that everyone knows. For some, it influences their whole lives. We speak, of course, of the constellations: the stuff of mythology, horoscopes, and all that. Astrology's influence upon human history has been as great as that of any other idea, and the affairs of some nations are still significantly influenced by astral projections. The reasons for the rise of astrology in the ancient world are not known with certainty, and probably differ from one civilization to another. The Egyptians believed that the stars were another world where our spirits rested after death. The design and arrangements of the Pyramids were closely correlated with the positions of stars in nearby star fields, in an attempt to recreate the ground-plan of the afterlife here on Earth. Since the celestial motions control the daylight hours, the tides, and the seasons, it is not altogether unnatural to believe that they control everything else as well. Such superstitions about the patterns of the stars have persisted for many thousands of years. People seem to have a natural inclination to believe that the course of their lives is determined by outside forces, and to identify unseen patterns behind the appearances. Yet, those same pictures that the ancients projected upon the sky to help them identify special groups of stars served a practical purpose. The slow variation in the appearance of the sky enabled long-lived civilizations to keep track of time in sophisticated ways. More important still, on a day-to-day basis, was the use of the night sky as a navigational aid. This is essential for seafaring nations. Whereas overland travellers can safely stop when the light fails and landmarks become invisible, sailors cannot.

Many peculiar myths are simply mnemonics for identifying the arrangement of particular groups of stars. The constellations have names that were picked out by other ancient cultures, who attached their own images to them. Today, we would no doubt make different choices (see Figure 4.18). But where did the original constellations come from? Who created this cornucopia on the dark night sky? When did they do it? And why? Ironically, in answering these questions, we shall discover that the constellations can tell us something about the past, even if they cannot foretell the future.

We can pin down when, and where, the constellation-makers lived by recalling that the Earth's axis precesses as it spins, like a wobbling top, so that the polar axis traces out a circle on the sky every 26 000 years. If we consider an observer on Earth, situated as in Figure 4.19, then the observer's horizon divides the sky in half. Only the part of the sky above the horizon is visible at any moment. If the latitude of the observer is L degrees north, then the North Celestial Pole lies L degrees above his horizon, and the South Pole lies L degrees below it. The rotation of the Earth makes the sky appear to rotate in a westerly direction around the North Celestial Pole. Stars rise at a point on the easterly horizon, then travel up the sky before reaching their highest point, after which they descend to set on the westerly horizon. Most stars follow this pattern, with seasons of visibility followed by seasons of invisibility. From Britain and from much of northern Europe, for instance, we can see Orion and Sirius in the winter, but not during the summer.

There are two groups of stars that are not seen to follow this pattern of nightly rising and setting. Stars within a circle that extends L degrees from the North Celestial Pole never disappear below the horizon. They can always be seen if the sky is clear. They are called the northern *circumpolar stars*. For European observers, they include the Great Bear and Cassiopeia. By the same token, there is a group of southern circumpolar stars within a circular region of the same angular extent around the South Pole. They are never seen by the observer in our picture, because they never rise above his horizon. Thus, the Southern Cross cannot be seen from northern Europe, even though it is visible in Tasmania. The size of these ever-visible and never-visible regions, and hence the number of different stars that are encompassed by them, varies with the latitude of the observer. The larger the latitude, the larger are the circumpolar regions of the sky, as can be seen from Figure 4.19.

The annual path of the Sun can be superimposed upon this picture. As we have already seen, when discussing the seasons, the Earth's axis of rotation is tilted with respect to the plane of its orbit around the Sun. So, from a terrestrial

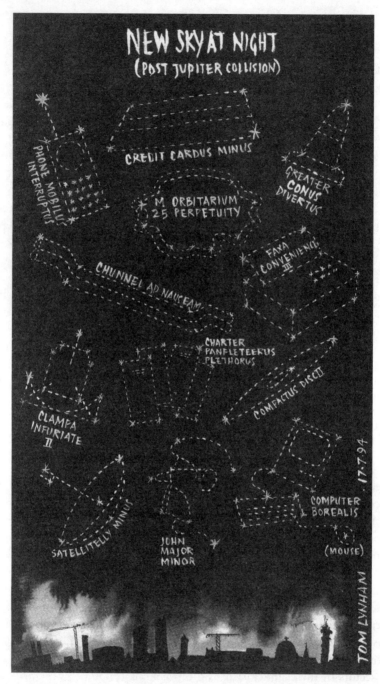

4.18 A contemporary version of the constellations by Tom Lynham, from the *Observer*.

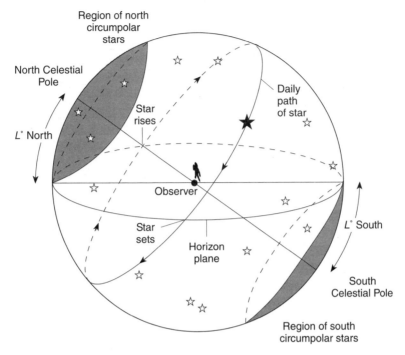

4.19 The celestial sphere. The observer is located at a latitude of L degrees north. Only one-half of the celestial sphere is visible to the observer at any one moment. Some stars are so close to the North Celestial Pole that they never disappear below the horizon. These are the circumpolar stars. A second group of stars, called the south circumpolar stars, are never seen by the observer because they do not rise above the horizon. See also Figure 4.13 on p. 169.

perspective, the Sun traces out on the celestial sphere a great circle that is inclined to the celestial equator, as shown in Figure 4.20. This path round the celestial sphere was in ancient times divided into the twelve signs, or 'houses', of the zodiac by the twelve constellations through which the Sun passed in sequence on its annual journey around the Earth. (We recall that, in ancient times, the Earth was believed to be the centre of the solar system.) These twelve signs are still used today in the astrological columns of popular news-papers. In fact, the signs of the zodiac differ from the constellations of the zodiac, even though they share similar names. Constellations are prominent groups of stars that have a noticeable shape. They are of different sizes, and contain different numbers of stars. The signs of the zodiac, by contrast, are equal sectors of the ecliptic: each of the twelve signs covers a zone that is 30 degrees long (so the total, 360 degrees, covers the whole circle), and by convention they are taken to be 18 degrees wide. At first, there was clearly a rough

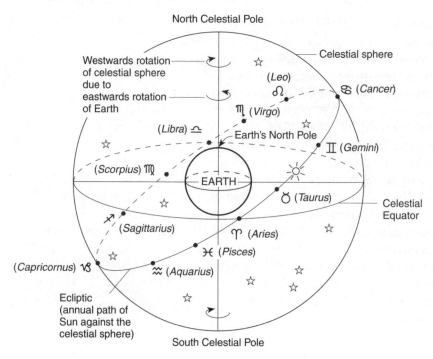

4.20 The path of the Sun on the celestial sphere. Its annual path is a great circle called the ecliptic. The ancients divided the ecliptic into twelve constellations, the so-called houses of the zodiac, in order to chart the path of the Sun.

correspondence between each sign and the constellation bearing the same name. But there were many more traditional ancient constellations than there were signs of the zodiac, and one might speculate that we see evidence here of two threads of invention that eventually became entwined. Although astrological purposes could be served by a neat twelvefold division, navigational needs might be less predictable and change with time. In this way, additions to the astrological scheme might both be necessary and persistent, once made.

The direction of the Celestial Pole slowly changes, tracing out a circle of angular radius 23.5 degrees in the sky every 26 000 years. As explained above, the direction of this Pole is apparent as the axis around which all the stars turn. Hence, in the distant past, sky-watchers would have seen a different direction from the one that we now see as the centre of the turning stars. If one finds an ancient record of detailed observations of the sky, the date of its authorship may be gauged approximately by noting which star is used as the nearest indicator of the North Celestial Pole. Today, that star is Polaris, but in 3000 BC it would have been Alpha Draconis. Knowing these features of the changing night sky, astronomers have sought to discover who 'created' the ancient constellations. The method is simple. If we inspect the ancient pattern of named constellations in the southern and northern skies, shown in Figure 4.21, then we find that there is a region of the southern sky that is empty of ancient constellations. Modern star-charts show that this region has been filled in by additions made during past centuries.* (Two beautiful hand-coloured maps of the Christianized medieval sky, drawn by Andreas Cellarius in 1660, are displayed in Plates 18 and 19.) Looking again at Figure 4.19, we see that this state of affairs is to be expected. An observer at latitude L degrees north cannot see a circumpolar disc of stars of angular diameter $2L$ degrees, centred upon the South Celestial Pole. Thus the size of the region devoid of ancient constellations tells us something about the latitude at which the constellation-makers lived. The diameter of the empty zone on the sky subtends about 72 degrees; so the constellation-makers should be found near a latitude of 36 degrees north. We can date them as well. The empty region is not centred around the present South Celestial Pole, nor would we expect it to be: the slow precession of the direction of the Polar axis on the sky rotates it on a 26 000-year cycle. We would expect the empty zone to be centred upon the direction of the Celestial Pole at the time when the constellation-makers were observing. It is found that the centre of the empty zone coincides with the position of the South Celestial Pole in the period 2500–1800 BC (Figure 4.22). One question now remains: who were they?

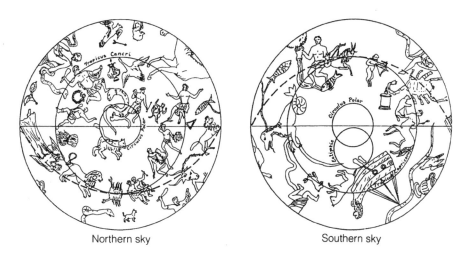

Northern sky   Southern sky

4.21 The ancient constellations of the northern and southern skies.

* For example, Hydrus the water serpent and Chamaeleon the chameleon were both devised by two sixteenth-century Dutch navigators, Frederick de Houtman and Pieter Dirkszoon Keyser, in order to fill the vacant space in the sky near the South Celestial Pole.

4.22 The zone of the southern sky that is empty of ancient constellations. N marks the present location of the South Celestial Pole; H marks its position at the time when Hipparchus observed the sky (190–120 BC); C marks the centre of the circular zone of the sky (marked 36°) that is empty of constellations when viewed from a latitude of 36 degrees north. The circular zone (marked 31°) demarcates the region of sky that Hipparchus could not have seen, assuming he observed from Alexandria (latitude 31 degrees north). The segments cut out by the intersection of the two circles give the regions of sky that the constellation-makers saw, but Hipparchus could not have, and vice versa.

Intriguingly, there is a further body of literary evidence which Edward Maunder, in 1909, and then more recently Michael Ovenden, in 1965, used to narrow down the candidate civilizations more specifically.* The earliest complete description of the ancient constellations, excluding the exact positions of the individual stars, is to be found in a poem by Aratus of Soli entitled *The*

* A Swedish amateur astronomer and civil servant, Carl Gottlieb Swartz (1757–1824), who studied at Uppsala University, pursued the problem of the origin of the constellations in a systematic way more than a hundred years before Maunder. In 1807 Swartz published his ideas in the book *Recherches sur l'origine et le signification des Constellations de la Sphère greque,* translated

Phaenomena ('The Appearances'), published in about 270 BC. His list of the constellations corresponds, in almost every respect, to the 48 listed by the great astronomer Ptolemy, together with their positions, in his catalogue of AD 137. Intriguingly, we know that St Paul, who like Aratus was a native of Cilicia, was familiar with this information, because he quotes the opening verses of Aratus' poem in his address to the Athenian Court of the Areopagus on Mars' Hill, which is recorded* in the New Testament (Acts 17). Aratus was educated in Athens, and his work would have been well known to an educated audience. Quotation boosted Paul's credibility by displaying his knowledge of Greek literature; and its particular content provided a sympathetic point at which to begin his sermon about the identity of their 'unknown god'.

The constellations of which Aratus writes were not his creation. For his poem was written as a tribute to Eudoxus. It versified Eudoxus' famous account of the stars, also entitled the *Phaenomena*, which had been written more than a hundred years earlier. And, in fact, judging by other passing references to particular constellations in their literature,† the Greeks were familiar with the constellations at least a thousand years before Christ.

Eudoxus of Cnidus lived between 409 and 356 BC, and was one of the greatest

---

from Swedish, which was then republished in a second edition under the shorter title *Le Zodiaque expliqué* in 1809. Swartz noticed the region of the southern night sky that was unpopulated by constellations and estimated its angular size to span about 40 degrees. He used this to date the epoch of the origin of the constellations at about 1400 BC and identified the coastal city of Baku, in Armenia on the Caspian Sea, at latitude 40 degrees north, as the most likely home of the society of seafarers and navigators who laid down the plan of the ancient constellations (see the intercept at approximately 50 degrees longitude on Figure 4.23). Swartz's maps of the ancient constellations with the 40-degree empty zone marked in the southern sky are shown in Figure 4.23.

* Aratus' poem begins:

> To God above we dedicate our song;
> To leave Him unadored, we never dare;
> For He is present in each busy throng,
> In every solemn gathering He is there.
> The sea is His; and His each crowded port;
> In every place our need of Him we feel;
> For we His offspring are.

St Paul's speech contains the words:

God that made the world and all things therein, seeing that he is Lord of heaven and earth, dwelleth not in temples made with hands; neither is worshipped with men's hands, as though he needed any thing, seeing he giveth to all life, and breath, and all things; and hath made of one blood all nations of men for to dwell on all the face of the earth, and hath determined the times before appointed, and the bounds of their habitation; that they should seek the Lord, if haply they might feel after him, and find him, though he be not far from every one of us: for in him we live, and move, and have our being; as certain also of your own poets have said, For we are also his offspring.

† In the fifth book of Homer's *Odyssey* we read that

> With beating heart Ulysses spread his sails:
> Placed at the helm he sat, and marked the skies,

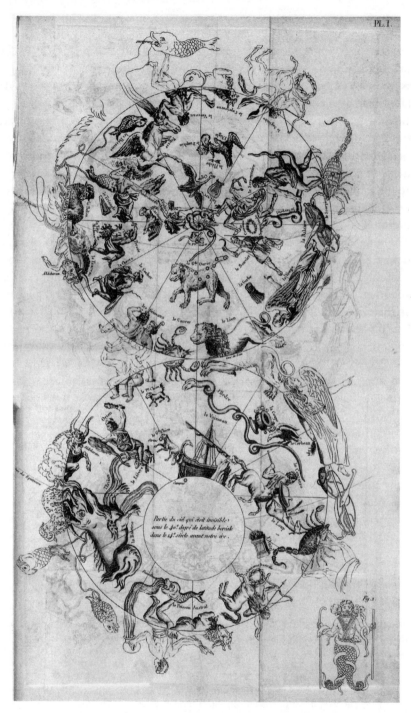

4.23 Carl Swartz's 1809 constellation maps showing the empty zone in the southern sky from which he deduced that the constellation-makers lived at 40 degrees of latitude in Baku.

mathematicians of the ancient world. He is best known as the author of the fifth book of Euclid's geometrical work, the *Elements*. He was lured into the study of astronomy by Plato's challenge to mathematicians to explain the ordered celestial motions. Besides his two important astronomical treatises on the appearance of the heavens, he is also famous for leaving his successors an engraved globe, 'the sphere of Eudoxus'. He used this for astronomical study and probably had the ecliptic, the Equator, known stars, and the names of some constellations marked on it. It must have been the prototype for the modern celestial globe that astronomers use to represent the information contained in Figure 4.19 in three-dimensional form. Unfortunately, neither his writings nor his sphere survive. We know much about them, however, from Aratus' poem, which was commissioned in 270 BC by King Antigonus Gonatus of Macedonia as a posthumous tribute to Eudoxus. The author's brief was to produce a tribute in verse, incorporating the astronomical content of Eudoxus' study of the heavens. Since the author was no astronomer, he stays close to Eudoxus' original, and provides a very detailed constellation-by-constellation guide to the sky.

One hundred and fifty years later, Hipparchus of Rhodes, the greatest of Greek astronomers (he discovered the precession of the Earth's Polar axis) studied Aratus' poem. He was puzzled by what he found. Neither Aratus nor Eudoxus could have seen the arrangement of constellations recorded there. They described arrangements of stars that never appeared above the horizon at the times when they wrote. Moreover, there were other stars, obvious now to Hipparchus, of which Aratus made no mention at all. There is an explanation for these discrepancies. The constellations were first identified by astronomers long before Eudoxus. And, as a result of the precession of the Earth's axis of rotation, the sky that they saw was significantly different from that seen by Eudoxus, Aratus, and Hipparchus. Hipparchus might even have begun to uncover the phenomenon of the Earth's Polar precession by seeking to reconcile the data of Eudoxus, in Aratus' poem, with what he knew of the sky in his own era, although there is no direct evidence for this.

It is evident that by a careful analysis of the constellations included, and omitted, from Aratus' poem one might determine the epoch for which it provides a correct description of the sky. In 1965, the Scottish astronomer Michael Ovenden carried out this analysis of the astronomical descriptions in Aratus' work to deduce both the latitude and date of the original creators of the information in Aratus' poem about the constellations. (A colleague checked the

---

Nor closed in sleep his ever-watchful eyes.
There view'd the Pleiads and the Northern Team,
And Great Orion's more refulgent beam,
To which around the axle of the sky
The Bear, revolving, points his golden eye.

analysis.) Ovenden found a latitude between 34.5 and 37.5 degrees north, and an epoch between 3400 and 1800 BC. This agrees remarkably well with the earlier deductions drawn from the absence of ancient southern constellations (2500–1800 BC), and offers confirmation of the idea that the original constellation-makers all lived at one epoch in one region. They predated Eudoxus by thousands of years. Eudoxus must merely have repeated the information he inherited from them without checking it against observations. If he had done so, he would have discovered that it described star patterns that were not visible to him, and omitted others that were. Aratus did the same, but could hardly be faulted—after all, he didn't claim to be an astronomer.

In 1984, Ovenden's Glasgow colleague, Archie Roy, carried out a more detailed study of the astronomical epoch to which Aratus' poem refers by using the detailed statements in the poem to deduce how the Tropics of Cancer and Capricorn, and the Equator, intersect the constellations. To appreciate the detail that makes this type of analysis possible, consider the information in the poem about the Equator (which is identified in the first three lines); Aratus gives a detailed specification of the associated constellations:

> In the midst of both, vast as the *Milky Way*,
> A circle trends 'neath earth like one in twain;
> And on it twice are equal days and nights,
> At summer's close and when the spring begins.
> As mark there lies the *Ram*, and the *Bull's* knees;
> The *Ram* along the circle stretched at length,
> But the *Bull's* crouching legs alone appear.
> And on it the bright *Orion's* belt,
> The *Water-serpent's* gleaming bend; The *Bowl*
> But small, the *Crow*, some few stars of the *Claws*;
> The *Serpent-holder's* knees are in it borne.
> It does not share the *Eagle*, messenger
> Of might who flies nigh to the throne of Zeus.
> On it the *Horse's* head and neck revolve.

Roy took this passage, together with two others that deal with the intersections of the Tropics of Cancer and Capricorn, and used the information to program a planetarium to recreate the appearances of night skies between the present and 5000 BC. There is a striking convergence of all the statements with the appearance of the sky in the Mediterranean latitudes of interest, as it would have been observed between about 2200 BC and 1800 BC.

We have followed three different lines of enquiry that point to the same location and time-frame for the constellation-makers. Clearly, Eudoxus could not have devised the famous sphere that bears his name, and from which the positions of the stars in Aratus' poem were ultimately derived. The astronomy embodied in his sphere, and perhaps the sphere itself, must have been inherited from another civilization whose astronomers were active more than 1500 years before Eudoxus was born. That sphere was probably inscribed to allow a navigator to use the constellations to set a course by remembering the order in which they would rise and set on the horizon. This would have been particularly useful because, unlike today's mariners, they would have lacked a convenient Pole Star to guide them.

One interpretation of this remarkable body of evidence is that the ancient constellation-makers created the astrological and mythological pageant on the sky as an embodiment of their own familiar spirits, heroes, and demons, and organized its layout in a comprehensive and memorable fashion for their own navigational purposes. Aratus' poem is threaded with concerns about peril at sea, and this implies that the originators of its astronomy were a race of seafarers, who required an understanding of the sky for navigational purposes. They may have been the inventors of the constellations, or they may have adapted a more primitive mythological scheme of star names into a system of practical use to navigators. There is a tradition that Eudoxus obtained his sphere, or the information needed to construct it, during his travels in Egypt, but nothing similar has ever been found in the vast collection of remains of ancient Egyptian civilization. Even so, if Eudoxus was given it during his own lifetime, why did the Egyptians give him information about the sky that was thousands of years out of date? They themselves could not have seen those sky patterns. Were they aware that they were giving him an inferior, grossly outdated model? If so, where is the evidence for the better ones that they were using? It is more likely that they too had inherited something they did not fully understand. Even if they knew it did not describe the sky that they could see, they were unable to correct it by further observations. So why did they not get a replacement from their original suppliers? To offer some possible answers to these questions, we need to narrow down the list of candidates for the first constellation-makers and users.

Let us leave astronomy, and turn to geography and history. The 36-degree line of latitude, which we identified as the home of the constellation-makers, runs through the Mediterranean and the Near East (see Figure 4.24). There were several advanced ancient civilizations nearby that might have framed the ancient constellations as navigational aids. The Phoenicians, living in the region now called Lebanon, can be discounted; despite their history as traders and seafarers, their civilization was in its golden period over a thousand years after the 2500 BC epoch that we are interested in. By contrast, although the ancient Egyptians were outstanding in their mathematical and technical achievements at that time, their latitude lies below 32 degrees north; this seems too far south for them to have been the constellation-makers. The Babylonians are surely better candidates. They have left thousands of cuneiform tablets detailing sophisticated

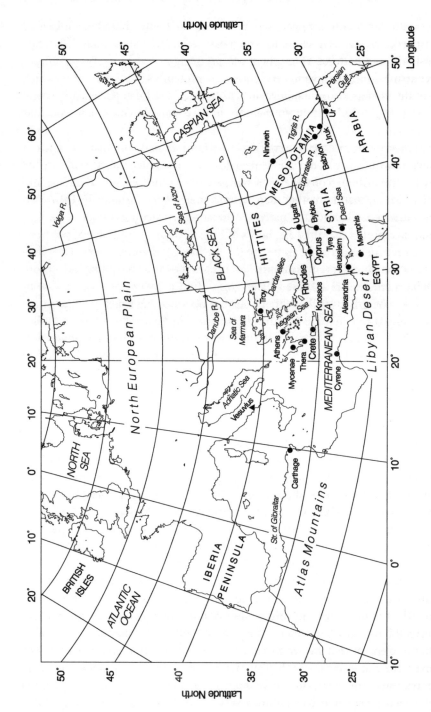

4.24 Lines of latitude running through the ancient Near East.

mathematical and astronomical studies that date back to 3000 BC. Moreover, their astrological interests were intensive. They had a special interest in recording the positions and patterns of the stars, because they believed that human affairs were controlled by them; we have already seen how their concern for the hierarchy of the planets informed the structure of their seven-day astrological week. Their written records give copious information about the stars, and associate some of them with the same images that we use today. The statements in Aratus' poem appear to be completely consistent with the sky as seen from the latitude of Babylon around 2500 BC. There seems little doubt that the subjects represented by the ancient constellations were deeply embedded in the Sumerian culture of Mesopotamia by that date. A further pointer is contained within the names of the constellations themselves. Astronomical tablets, dating from about 600 BC, have been discovered in the vicinity of the River Euphrates. They give Greek names for the constellations, but the images the star patterns represent seem to be much older. For example, the constellation that we still call Taurus, 'the Bull', is referred to in these tablets as the 'Bull-in-front'. In those times, the year was measured from the start of spring, which was defined by the vernal equinox (the day when there are equal hours of daylight and darkness). This, like the autumnal equinox, occurs when the ecliptic intersects the projection of the Earth's Equator on the sky.

At present, the Sun is in the constellation of Pisces at the vernal equinox, but at the time of Hipparchus it was in Aries, and Ptolemy makes Aries the first constellation of the zodiac. The description 'Bull-in-front' indicates that the name of the Bull was given to the constellation when it was at the front of the year—at the time of the vernal equinox. If we calculate when the Sun was in the constellation of Taurus during the vernal equinox, we get 2450 BC: nearly two thousand years before the tablets were inscribed, and in remarkable accord with our other indicators about the origin of the constellations. Moreover, at this early epoch there is a logic to the rest of the sky: at the summer solstice the Sun was near Regulus, the brightest star in Leo; at the autumn equinox it was near Antares, the brightest star in Scorpio; and at the winter solstice it was close to Formalhaut, the brightest star near Aquarius, the water-pourer.

This is not quite the end of the story. Although the Babylonians may have been the original shapers of the constellations, their seagoing activities seem to be too little, and in the wrong latitudes, to make great use of the elaborate system of constellations described by Eudoxus and Aratus. This dissonance led Roy to seek out another ancient seafaring civilization, which might have taken up and improved the Sumerians' astrological system for navigational use around the Mediterranean. There is only one candidate civilization at the right latitude (roughly 36 degrees); it is to be found less than a thousand miles west of Babylon, on the island of Crete—the home of the Minoans.

Until the beginning of the twentieth century, the Kingdom of Minos meant little more than the lost land of Atlantis: the home of mythical figures like Daedalus and Icarus, or of the great Minotaur, half bull and half human, prowling the Labyrinth. Then, gradually, archaeologists began to substantiate previous suspicions of a great innovative culture, centred upon Crete. References to trading activities between Crete and other Mesopotamian cultures can be found as early as 2350 BC; their trade with Egypt was extensive, and treasures of Egyptian origin have been found among the ruins of the Minoan palace at Knossos. The range of non-indigenous building materials that they used gives an impression of extensive seafaring throughout the entire Mediterranean region. But, at its climax, this sophisticated culture came to a sudden and catastrophic end. In about 1450 BC, a great natural disaster destroyed their entire civilization in one fell swoop. They had weathered an earlier earthquake around 1700 BC, but the disaster that followed appears to have been of a different magnitude. A huge volcanic eruption occurred in the Aegean at Thera in those times, and an explosion took place that left a crater hundreds of metres deep, encompassing nearly a hundred square kilometres. The ash, debris, ground tremors, and huge waves that resulted simply eliminated the Minoans. Their old harbours show evidence of dramatic compaction and movement of stone. What was not destroyed fell prey to other invaders; suddenly, the most advanced European civilization of its time was gone.

No documents or astronomical devices have been found in the ruins of Minos to prove that the Minoans were the great constellation-users and navigators around whom the sky turned in the third millennium BC. But they certainly fit the bill. Their trading horizons were growing far and wide in 2500 BC; they lived on the 36th parallel of latitude; their navigational and constructional skills give the impression of being able to adapt and supersede things that they learned from other cultures. They had strong trading links with Babylon, and would have been exposed to their astrological pattern of the constellations. Roy speculates that the source of the celestial globe that Eudoxus found in Egypt, with its mysterious fossilized picture of the heavens as they could only have been seen two thousand years before, was Minos. If so, the reason why it was never replaced by an updated version becomes clear. In the period between 2500 BC and the time of Eudoxus' visit, more than two thousand years later, the Minoan civilization had been utterly destroyed. And of their star-finding, nothing but the story of Eudoxus remains.

Even if this story provides the explanation for the overall layout and sky-coverage of the ancient constellations, there are still many possibilities regarding the development of the different constellations, whether they arose at the same time, or over an extended period. Alex Gurstein, a Russian historian of ancient astronomy, has sought to explain the appearance of particular constellations at much earlier times by considering their place as markers of key astronomical features of the sky. These marker-points change over thousands of years due to the precession of the Earth's rotation axis, and so new constellations get defined as markers in different millennia. There is no suggestion that these ancient sky watchers needed to understand the phenomenon of precession. They probably attributed the lack of named groups of stars at special points on the sky to oversights by previous generations, or perhaps even to great changes in the heavens brought about by the will of the sky gods.

Gurstein proposes that astronomical observations of the Sun's movement along the ecliptic—the so-called *via Solis*—would have established a correlation between the appearance of the night sky and the seasons of the year. This would naturally lead to the identification of four special groups of stars, one for each season. The seasonal changes are marked by the vernal equinox, the point of the summer solstice (when the Sun is highest in the sky at noon), the autumn equinox, and the point of the winter solstice (when the Sun is lowest in the sky at noon). They would have been appreciated when it became evident that the annual motion of the Sun on the sky allows the seasonal changes to be predicted reliably. Gurstein believes that the identification of the first constellations was primarily to mark important areas of the celestial sphere rather than simply uniting groups of bright stars for symbolic reasons or for navigation. The 26 000-year precession of the Earth will cause the position of the markers of the four seasons to change over thousands of years and will require new marker constellations to be introduced. The plane of the ecliptic remains virtually unchanged on the sky in the meantime. The marker constellations therefore move anti-clockwise through the signs of the zodiac (which means literally 'circle of animals'), passing a sign every 26 000 ÷ 12 = 2140 years. Therefore the same marker stars will define the equinoxes and solstices reasonably invariantly for about 2000 years.

Epoch	Spring	Summer	Autumn	Winter
8–7000 BC	Cancer	Libra	Capricorn	Aries
6–5000 BC	Gemini	Virgo	Sagittarius	Pisces
4–3000 BC	Taurus	Leo	Scorpius	Aquarius
2–1000 BC	Aries	Cancer	Libra	Capricorn
1–2000 AD	Pisces	Gemini	Virgo	Sagittarius
3–4000 AD	Aquarius	Taurus	Leo	Scorpius

Gurstein investigated the particular religious and mythological symbols that were prevalent in known societies at different epochs and would have led to the choices of creatures to signify the star markers. A clue to the chronology is also likely to come from the sizes of the constellations on the sky. The largest will tend to be the first ones to be picked out as markers. He concludes that the first four constellations on the path of the Sun were picked out during the sixth millennium BC, possibly within the region of the Earth that spread the Indo-European culture and languages.

▓▓ Study in scarlet: the sources of colour vision

I'm afraid of the dark, and suspicious of the light.

Woody Allen

In the second chapter, we looked at some of the restrictions that habitability imposes upon a celestial body. Two properties emerged as important for the evolution and maintenance of atom-based life on a solid, stable, planetary surface: the existence of a stable 'main-sequence' star like the Sun, and the presence of a gaseous atmosphere. A third property, a rotation of the planet upon its axis, is very likely; an unlikely coincidence of circumstances would be necessary to prevent it. We would expect these to be features of planets where the spontaneous evolution of life is probable. But these properties combine to create a property of the resulting planetary environment to which an adaptation can occur that is as unexpected as it is far-reaching.

The mixture of wavelengths emitted by a stable star like the Sun; the daily alternation of periods of light and darkness that arise because of planetary rotation; and the scattering and absorption of the star's light by a planetary atmosphere: these processes combine to create conditions of illumination on the surface of the planet that make the evolution of a particular type of colour vision advantageous and adaptive.

If we consider the reception of scattered sunlight on the Earth's surface, we know that much of the Sun's radiant energy is absorbed by water vapour and ozone in the atmosphere. The Sun's intensity of emission has its peak in the blue–green portion of the colour spectrum (see Figure 4.25), but the scattering of light by molecules in the Earth's atmosphere affects the shorter wavelengths (indigo, blue, and green) most; so they do not reach our eyes, thus the disc of the Sun appears yellow. The scattered blue light is what makes the rest of the sky blue. Pure water appears blue for the same reason. If we look away from the Sun, we are seeing light that has been scattered in the atmosphere. The short-wavelength (bluer) photons are scattered most, and hence the sky is blue; if we look towards the setting sun (see Plate 10), we receive the long-wavelength (redder) photons that are scattered least *en route* to our eyes. (Ironically, the most spectacular sunsets, with vivid reds, oranges, and purples, occur over the most polluted industrial cities or in the vicinity of volcanic eruptions—because

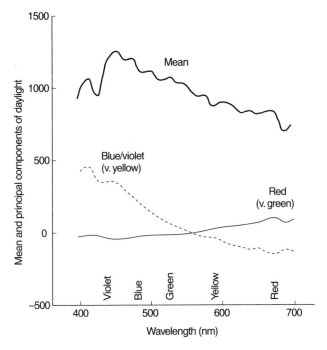

4.25 Relative spectral intensities of the mean, the blue–yellow and the red–green components of daylight.

the air contains a superabundance of car exhaust gases or smoke particles which enhance the scattering process.) When the scattering particles in the atmosphere become larger—water-vapour droplets, snowflakes, or particles of sand or dust—then the scattering ceases to depend significantly upon the wavelength (colour) of the sunlight.* All wavelengths are then scattered more or less equally, and the result is a white or misty scene. This is why clouds and overcast or misty skies appear white, and why the ocean appears white when viewed through the spray from a sandy beach in rough and windy conditions. There are also some white animals, like polar bears, that owe their appearance to this effect rather than to the presence of an intrinsic white-coloured pigment. The shafts of fur on a polar bear contain tiny bubbles of air that scatter incident light and give the collection of transparent hairs a white appearance.

Moonlight, because it is just sunlight reflected off the face of the Moon, has a very similar spectrum to direct sunlight, although its intensity is a million times

---

* The intensity of scattered light is proportional to the fourth power of its frequency. Hence, over the range to which the eye is sensitive, the intensity of blue light will be 16 times greater than that of red light (which has twice its frequency).

lower. The total integrated starlight from the rest of the Universe is a thousand times fainter still. In between moonlight and sunlight, we have twilight. Its colour spectrum differs from that of sunlight and moonlight; all three are shown in Figure 4.26.

At twilight, rays of sunlight must pass through more of the Earth's atmosphere before reaching us, and the absorption of yellow and orange light by ozone molecules becomes important. This gives the sky colour a slight tinge of magenta in the last 30 minutes before sunset, and in the 15 minutes before sunrise.

We have mentioned the transient twilight phenomenon because it may be the reason for a peculiar feature of human colour vision. In 1819, a Czech physiologist, Jan Purkinje, noticed a curious phenomenon as he watched the flowers in his garden at twilight. He realized that the relative brightnesses of

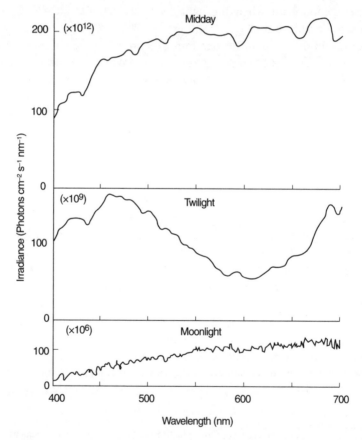

4.26 The spectral composition of moonlight, sunlight, and twilight. Data taken in the summer of 1970 at Einewetok Atoll.

differently coloured flowers were changing as the light faded. Red flowers became black, while green leaves remained green and bright. At low light levels, the human eye becomes more sensitive to blue and green light than to red light (Figure 4.27).

At first, this behaviour seems to be maladaptive because, as one can see from Figure 4.26, moonlight (and also starlight) contains more long-wavelength (red) light than daylight. We might therefore have expected human sensitivity to red light to increase, not decrease, at low illumination levels. If, however, we compare Figures 4.26 and 4.27, we see that, when light levels fall, the wavelength at which the eye is most sensitive shifts to where the greatest sensitivity is required in twilight conditions.* The implication is that this twilight zone is the most dangerous one: lighting conditions are varying quickly; nocturnal predators are appearing; and fatigue is setting in. It might well be more adaptive to have better vision during this brief, but dangerous period than to optimize reception to the spectrum of moonlight when the light levels are too low to allow any real advantage to be had from it.

The cross-cultural human descriptions of colours are intriguing. We know that colour is determined by the wavelength of light, and the spectrum is completely continuous between red and violet. Nevertheless, we all identify a small collection of definite colours—red, orange, yellow, green, blue, indigo, violet—

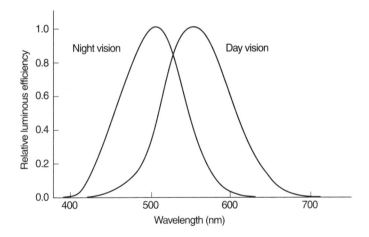

4.27 The efficiencies of human colour reception in day and night vision.

* However, the peak is a fairly shallow one and there is still much dispute as to the reason for the 'blue-shifting' of the eye's sensitivity, a feature shared by large numbers of apparently unrelated crustaceans and vertebrates. Perhaps there is some biochemical constraint upon the molecules involved, or adaption to a past environmental feature that no longer exists, or some as yet undiscovered difference between the rods and cones (light-sensitive cells) of the eye.

and exaggerate the differences between them.* There have been detailed studies of colour words—and the hues to which they correspond—used in diverse cultures and languages. A study of 98 languages in which native speakers were shown a range of different colour cards found that there was virtually a universal choice of the parts of the light spectrum to which colour words were attributed. The principal difference was in the number of colours distinguished by colour words. Here, there was also a general trend. The simplest languages had words only for black and white; the next commonest addition was red, followed by green and yellow with roughly equal frequency, followed by the addition of blue, then brown, and then purple, pink, orange, and grey. The pattern of occurring colour words is shown in Table 4.4; only 22 out of the 2048 logically possible sets of the eleven basic colour terms were found in the languages studied. These studies have been interpreted as indicating the way in which our colour lexicon tends to develop. A pattern of evolutionary development of colour words is suggested, as shown in Figure 4.28. Although the trend is clear and not entirely surprising, care must be taken not to stretch this rather one-dimensional data too far. Keeping a list of number words divorced from the situations and circumstances in which their speakers exist is fraught with potential biases. Live in the snow and you will have need for a different spectrum of colour words than if you live year-round under an azure sky or wander in verdant forests.

Black and white are the first terms needed in order to convey information about the levels of light and darkness in the environment. The next most complicated vocabularies add terms for 'red', which includes shades of brown, and is often linked to the description of soil, or of blood. Even today, we recognize the prevalence of black, red, and white as symbols of office, and they are frequently used for uniforms or ceremonial dress: recall Stendahl's *The Red and the Black*.

Our colour categories do not seem to be accidental. They are linked to the fact that the visual system is three-dimensional. In bright conditions, the eye has three types of detector (cones) in the retina, with photochemical pigments whose peak sensitivities are tuned respectively to the long, middle, and short wavelength regions of the visible spectrum. The eye registers three separate pieces of information, which are then weighted and combined to produce our

* The seven colours of the spectrum that Newton picked out have an interesting history. In his first lectures and writings on colour in 1669, Newton delineated only five primary colours: red, yellow, green, blue, and purple. Later, in 1671, he introduced further secondary colours. Orange and indigo seem to have been added so as to bring the number up to seven, because he believed that light vibrations were analogous to sound vibrations and so the number of primary colours should correspond to the seven musical tones of the diatonic scale. The choice of indigo as a distinctive spectral hue no doubt owes something to its commercial prominence in Newton's day. Indian dye (= indigo) was introduced into Europe during the sixteenth century and was widely used thereafter. Today, most scientists encounter the term 'indigo' only in a listing of the colours of the spectrum.

Table 4.4 The 22 colour vocabularies originally identified by Berlin and Kay in their study of traditional peoples. The simplest (type 1) have only two colour words, for black and white; the most sophisticated (type 22) have eleven distinct colour words.

No. of basic colour terms	Perceptual categories encoded in the basic colour terms										
	White	Black	Red	Green	Yellow	Blue	Brown	Pink	Purple	Orange	Grey
2	+	+	–	–	–	–	–	–	–	–	–
3	+	+	+	–	–	–	–	–	–	–	–
4	+	+	+	+	–	–	–	–	–	–	–
4	+	+	+	–	+	–	–	–	–	–	–
5	+	+	+	+	+	–	–	–	–	–	–
6	+	+	+	+	+	+	–	–	–	–	–
7	+	+	+	+	+	+	+	–	–	–	–
8	+	+	+	+	+	+	+	–	+	–	–
8	+	+	+	+	+	+	+	–	–	–	+
8	+	+	+	+	+	+	+	–	–	+	–
9	+	+	+	+	+	+	+	–	+	–	+
9	+	+	+	+	+	+	+	–	+	–	+
9	+	+	+	+	+	+	+	–	+	+	–
9	+	+	+	+	+	+	+	–	–	+	+
10	+	+	+	+	+	+	+	+	+	+	–
10	+	+	+	+	+	+	+	+	+	–	+
10	+	+	+	+	+	+	+	+	–	+	+
10	+	+	+	+	+	+	+	–	+	+	+
11	+	+	+	+	+	+	+	+	+	+	+

$$\begin{bmatrix} \text{White} \\ \text{Black} \end{bmatrix} \longrightarrow [\text{Red}] \begin{array}{c} \nearrow [\text{Green}] \longrightarrow [\text{Yellow}] \\ \searrow [\text{Yellow}] \longrightarrow [\text{Green}] \end{array} \begin{array}{c} \searrow \\ \nearrow \end{array} [\text{Blue}] \longrightarrow [\text{Brown}] \longrightarrow \begin{bmatrix} \text{Purple} \\ \text{Pink} \\ \text{Orange} \\ \text{Grey} \end{bmatrix}$$

4.28 The evolutionary development of colour description suggested by the data in Table 4.4.

final sensation of colour. These three visual sensitivities can be re-expressed as the brightness level, the yellow–blue variation, and the red–green variation. They are sometimes represented in a colour circle (see Figure 4.29), first introduced by Isaac Newton in 1704. This circle joins the two ends of the spectrum in order to illustrate the human tendency to find long-wavelength red and short-wavelength violet more alike than other colours of the spectrum that are much closer to each other in wavelength.

It is now a challenge to identify aspects of the environment to which adaptation would tend to select for dark–light, blue–yellow, and red–green discrimination together with the psychophysiological association of the two extremes of the colour spectrum. The overall range of spectral sensitivity of the human eye (400–700 nanometres*) reflects the range of wavelengths of solar radiation that reach us after passing through the atmosphere. We might, therefore, wonder whether more detailed aspects of the transmitted and scattered light influence

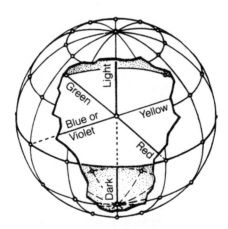

4.29 Newton's colour circle. A schematic three-dimensional parameterization of our mental representation of colour in terms of the axes of lightness v. darkness, blue v. yellow, and red v. green.

* A nanometre is one thousand millionth of a metre.

the fine details of colour reception. The dark–light sensitivity is necessary to accommodate the large variations in light levels that occur in natural environments because of shadow, cloud cover, the phases of the Moon, and the changing height of the Sun in the sky. We have already seen how the transition from light-adapted to dark-adapted vision at twilight points towards an adaptive structure. For the yellow–blue axis of colour contrast discrimination makes sense if it is an adaptation to the colours introduced into the environment by the Sun. The blue of the sky is a primary influence, while the centre of the rest of the solar spectrum (after the blues and violets have been subtracted out by scattering) is characteristic of direct sunlight and, like the face of the Sun, is yellow in colour. The blue–yellow variation mirrors the range of colours of sunlight: from direct, overhead sunlight to the scattered blue sunlight that colours the sky and water. The red–green variation in colour vision may also be linked to the influence of atmospheric scattering. The red portion of sunlight, although scattered least by molecules of air, is the part that is most readily absorbed by water vapour along its path. Thus an increase in the water vapour content of the atmosphere produces a reduction in the red component of the sunlight reaching the Earth's surface when the Sun is low on the horizon. After the reds have been removed in this way, the central wavelength of the remaining light lies in the green. This linking of colour opposites, like blue and yellow, by the process of averaging what is left of the spectrum after subtracting part of it, also has the effect of creating a closed circle of colour variations, of the kind shown in Figure 4.29.

Atmospheric influences alone could thus have begun a sequence of adaptations because of the selective advantage conferred by genes that promote the development of neural processing for distinguishing, simultaneously and economically, the three colour variations.

There are further environmental influences that reinforce adaptive responses to particular colours. The greens of leaves are produced by chlorophyll.* Birds

---

* In environments where there is a significant seasonal decrease in the hours of sunlight, or in its intensity, the leaves of deciduous trees will turn brown to produce the spectacular mixture of browns, reds, and yellows that in America is called 'fall' and in England 'autumn'. After midsummer, large-leaved trees, like oaks, invest more nutrients in their trunk and root systems than in sustaining leaves. When light levels are low and temperatures fall, the large leaves lose heat readily through their large surfaces, and cannot maintain a temperature high enough for chemical reactions to produce an adequate supply of nutrients. It is more economic for the tree to shed its leaves and grow a new set when the spring comes, rather than to use scarce resources in retaining them through the winter when in any case there is little light for them to gather.

Evergreens have a different strategy. Their leaves are small and needle-like, and present a smaller surface area from which heat loss can occur. They can thus maintain a useful level of chemical activity during the winter months. In this way, a spruce tree can afford to keep its needles all winter and take advantage of occasional bright spells of sunlight. In the summer, with only narrow needles, it is far less efficient at utilizing sunlight than the large-leaved oak, which has emerged from its winter hibernation furnished with a new set of large leaves.

and animals that forage will fare best searching for food sources that can be readily identified in their natural surroundings. Many plants will propagate most successfully if they are easily noticed, because they need insects to pollinate them, or they rely on ingestion and excretion by other living things to spread their seeds. There is scope here for the co-adaptation of these two propensities, to mutual advantage. The greens of plants are determined by chemistry; hence the most readily identifiable berries and fruits will be those with strong contrast colours—of which red is the most striking and the most common. Likewise, gatherers of such fruits will benefit from acute discrimination in the green, and in the red–green, contrast range (Plate 14). If coloured food sources are being exploited by, say, birds with colour vision, then later arrivals on the evolutionary scene (like primates), competing for the same resources, will find improved colour vision adaptive. Creatures that feed solely on grass, or on meat, tend to be colour-blind. Not unrelated to this use of colour vision might be the tendency of our visual system to bring colours like red into the foreground and push blue into the background. It is as if there is an adaptation to the ever-present back-drop of the sky and the advantage to be gained by seeing red first as a foreground highlight.

The array of colours on view in the natural world derive from chemical pigments and from the effects of scattered light. In some notable cases the colour that we see derives from a combination of them both. Light-scattering effects take three forms. Diffraction, when light passes through a small opening and past an opaque object, can be seen displayed by the colours of a spider's web hanging near a window. Interference of different waves of light is the source of colouring in the thin wings of a dragonfly and, most spectacularly, in the peacock's fan (see Plate 13). There, curiously, it is the melanin pigment in the barbules of the feathers that is reponsible for the optical interference pattern. The third, and most common, contribution of surface structure to natural colour is provided by the phenomenon of light scattering discovered by John Tyndall in 1869. The most dramatic example is the blueness of the sky, which Tyndall was the first to explain. Unlike the effects of interference and diffraction, the results of Tyndall scattering are not iridescent; that is, the colours seen do not vary with the angle at which they are viewed. Tyndall showed that the higher the frequency of light (that is, the bluer the colour), the more it is scattered by very small particles. This is why dry cigarette smoke has a blueish tinge and why some human eyes are blue. Minute protein particles in the iris of the eye perform the scattering of white light entering the eye. As you grow older these particles become slightly

---

When leaves are green in midsummer, the chlorophyll, responsible for the green colour, breaks down in the heat, but is being constantly replenished. In the autumn, the replenishment ceases and the other russet browns, formerly overshadowed by the brilliant greens, start to dominate.

larger and blue eyes will fade. Brown and yellow eye colours are produced by the presence of the pigment melanin which prevents the scattering (green arises right on the borderline where the yellow combines with the blue effect). Tyndall scattering is also responsible for the blue feathers of kingfishers and budgies, and for the blue tinge to the chin of a dark-haired man that can be seen after he has shaved. Remarkably, Tyndall scattering is also responsible for the coloration of most *green* bird feathers and the skins of many frogs and lizards (the chlorophyll pigment, which produces the green colours of plants, does not occur in animal tissues). A green tree frog (see Plate 16) is chemically yellow. But the yellow carotenoid pigment it contains acts as a filter for the scattered light, and the combination of yellow with the blue effect from Tyndall scattering makes the frog appear brilliant green. If you place a dead green frog in alcohol, the yellow pigment dissolves, and it will appear blue.

The carotenoid pigments which also colour the frog are responsible for the common yellows and oranges seen in plants, fish, and animals. The paradigm is provided by the carrot, after which these pigments are named, but their effects can also be seen in things as disparate as tomatoes, goldfish, and flamingos.

The most common pigment is black melanin, which colours things like human hair, skin, and blackbirds' feathers. It can also come in shades of brown, and generally provides the backdrop against which we see the more spectacular blues and greens produced by scattering.

The other common natural colours are reds and purples. Reds derive primarily from haemoglobin, or its compound oxyhaemoglobin, which colours the blood of humans and most animals. In the ears and nasal regions of cats it is responsible for the pinky fleshy colourings. In the raw it is displayed by the butcher's shop, where we can see it in the muscle cells of the meat chops more vividly than in blood. The more active a creature's existence, the more oxygen-carrying capacity it requires, and the redder its blood. Accordingly, deep-diving whales have very dark muscle coloration, whilst some rather inactive fishes actually have colourless blood. Finally, purples, together with some deep reds and blues, occur in plants because of a dissolved form of an anthocyanin pigment in their sap. This is the source of the familiar coloration in beetroot, rhubarb, and red grapes; and thereby, most impressively, of red wine.

We can identify four adaptive uses of colour in living things. First, it is used to attract attention: for example, flowers signal their presence to insects;* coloured fruits signal that they are good to eat (Plate 14). Second, it gives warning: for

---

* The distance at which a bee will turn towards a flower has been found to be proportional to the size of the flower head. Hence, small or isolated flowers need to be especially brightly coloured, with a high contrast against the green foliage, in order to gain reproductive advantage from not investing resources in larger blooms. Some impressive computer-generated flowers, created using simulations of real plant growth and decay, can be seen in Plate 17.

example, luridly coloured reptiles signal that they are poisonous (Plate 15). Third, it makes for the possibility of camouflage (Plate 12) or mimicry. Fourth, it acts as a stimulus to the emotions. Courtship displays make abundant use of colour signals (see Plate 13 and Plate 21*), and baboons display brilliantly coloured regions of their posteriors to indicate their sexual availability. As a result of this history, animals with colour vision respond differently to different colours. Monkeys prefer blue to green to yellow to orange to red; they usually have an aversion to red and orange, but are mildly attracted to blue and green.

One of the distinctive features of humans is their ability, and propensity, to colour themselves with artificial pigments and coloured objects. From war-paint to cosmetics, this tendency is an enduring human trait. It has many functions, which mirror the four we have just highlighted: the desire to be seen; to transmit information about rank and status, or warn of danger; to remain unseen; and to inspire admiration, respect, or fear. Some colours have become especially powerful arousers of the emotions. The prime example is the colour red, which, as we have already seen, is the first colour to be added to human vocabularies after black and white. It is also the commonest colour used by birds and flowers. Its effect upon humans is impressive: in cases of brain damage, red vision is the last part of colour vision to disappear, and the first to reappear if recovery occurs. But it is also puzzling. It signals danger, as in the eyes of the poisonous tree frog (Plate 16), and so is often used as a warning signal ('red for danger'), but it is also used cosmetically to heighten sexual attraction. Why does it have this confusing dual symbolism? While one can think of natural phenomena—like flames—of similar colour which send signals of safety and danger, it has been suggested by Nick Humphrey that it is the very ambiguity of our response that is most

---

* The orange-crested gardener bird (*Amblyornis subalaris*) is one of the rarest birds on Earth. About the size of a common starling, it lives only in a few dark, inaccessible mountain forests of New Guinea. Plate 21 was the result of the first observations of these birds constructing their bowers and performing their subsequent courtship ritual. This painting is based upon a collection of photographs taken in dark conditions over a period of many weeks by Heinz Sielmann. The order and detail of the bower is extraordinary. The central stem of the bower is constructed around a young sapling which has been surrounded by velvety moss. A central divide has been marked by carefully positioned yellow flowers, and two collections of objects arranged on each side of it. The left-hand side was decorated by embedding the moss with dozens of iridescent blue beetles, while the right-hand side was composed with pieces of blue snail shell. This part of the bower is like an exhibition of valuables to attract the attention of prospective mates. When a female passes by, the male (on the left in Plate 21) displays his orange crest and dances in front of his decorations. If he and all his works are impressive enough, she will eventually join the dance. The front of the bird's garden is meticulously delineated by rows of coloured fruits. The front border is fenced in by a network of tightly woven twigs which also provide a protective, waterproof, dome-shaped canopy. Needless to say, this work of art needs constant maintenance in the face of wind and rain, and the attentions of thieves. The result is one of the most striking creations in the whole animal world.

important. It seems to play the role of heightening our concentration in preparation to receive further information. The message that red sends depends upon the context, and we need to gather more information before the right conclusion is drawn. The very ambiguity of the situation, with the possibility of a totally incorrect response, causes the state of heightened attention that red so often stimulates.

The evolutionary adaptation to colour, and the strong responses that we have to it, mean that the artificial colours of our modern environment can be manipulated to produce particular responses. A striking example of unfamiliar colour signals from a familiar object is provided by the computer-generated animal shown in Plate 20. This is something that, whether consciously or not, plays a role in the choice of domestic decorations, the colour schemes of classrooms, hospitals, and other public buildings. Yet, for the most part, our environment is a haphazard mish-mash of many coloured objects. The effect is to dilute our sensitivity and response to colour symbols. Sensing this trend, Humphrey writes of the appearance of his study, and of the masculine tendency to neutralize colour information at the expense of other descriptors.

As I look around the room I'm working in, man-made colour shouts back at me from every surface: books, cushions, a rug on the floor, a coffee-cup, a box of staples—bright blues, reds, yellows, greens. There is as much colour here as in any tropical forest. Yet while almost every colour in the forest would be meaningful, here in my study almost nothing is. Colour anarchy has taken over. This has dulled our response to colour. From the first moment that a baby is given a string of multi-coloured—but otherwise identical—beads to play with he is unwittingly being taught to *ignore* colour as a signal.

Our tendency when teaching very young children is to get across the names of things, and the number of things; rarely do we place much emphasis upon their colours. If we consider how colour is used in Western artistic representation, it is striking that its use as a symbol was so restrained until the end of the 19th century. Other types of symbolism have been far more influential. Only with the development of abstract painting and other forms of modern art has the dramatic use of colour as a primary symbol become noticeable. One recalls Picasso's 'Blue Period', and the work of Mondrian, Vasarely, and Kandinsky, in which there is a strong appeal to our innate responses to particular colours. They are not being used simply to supply 'natural' colours to symbols pregnant with other meanings—as is the case with landscapes—or simply to reproduce the colours of natural objects—like fruits and flowers—to which we have innate responses. Rather, they reach out to touch a more basic instinctive reaction to colour. Wassily Kandinsky recognized how colour changes a person's mood and responses to pictures:

Colour is a power which directly influences the soul. Colour is the keyboard, the eyes are the hammers, the soul is the piano with many strings. The artist is the band which plays, touching one key or another to cause vibration in the soul.

The German Bauhaus school of design tried, in the 1920s, to develop a new form of iconography. Ludwig Hirschfeld-Mack, a long-term member of the school, tells* of one of their early studies, which investigated human propensities to ally shapes with particular colours:

A very interesting seminar was held during those early years. It was under the leadership of Paul Klee and Wassily Kandinsky and others. They sought to discover the reactions of individuals to certain proportions, linear and colour compositions . . . In order to find out whether there is a universal law of psychological relationship between form and colour, we sent out about a thousand postcards to a cross-section of the community asking them to fill in three elementary shapes, the triangle, square, and circle with three primary colours, red, yellow, and blue, using one colour only for each shape. The result was an overwhelming majority for yellow in the triangle, red in the square, and blue in the circle.

In Chapter 2 we saw something of Georges Seurat's use of point-like applications of colour to produce coloration and shadow with an intrinsic quality that is not meant to look as if it derives from the angle or intensity of sunlight. In fact Seurat had been influenced by the poet and scientist Charles Henry, who advocated links between emotional moods, colours, and the directions of lines in the composition. Seurat associated the three moods of gaiety, calm, and sadness with the primary colours red, yellow, and blue. Gaiety was also associated with rising lines; sadness with descending lines; while lateral lines were held to convey calmness and stasis. These recipes can be seen at work in a picture like *La Grande Jatte* (Plate 4).

There is much attention to shape and form in modern design; far less to the use of colour. But our innate colour sense is no less important than our instinct for pattern and order, or our desire for symbols of security. To use colour in ways that please requires an understanding of how it is used in Nature, and why, and how our visual sense evolved to accommodate its natural forms. Its presence is a gift of the sunlight; a byproduct of the need for habitable planets to orbit stars, to be cocooned by atmospheres, and to spend half their time with their backs turned upon their parent star. Without it, the monochrome world would be a bland and less inspiring place. Buried beneath layers of learning lurk our innate responses to colour. Occasionally, in moments of fright, or of wonder, they emerge uninvited from a repertoire that once melded us to this extraordinary environment of air and sky, of leaves and bright water, that bathes in the light of a star called the Sun.

* In *The Bauhaus* (Croydon, Australia, 1963).

Outward bound: the way of the world

Be humble for you are made of earth.
Be noble for you are made of stars.

Serbian proverb

One of the most interesting features of the pattern of progress in science is the way in which greater understanding of reality, and our increasing success in predicting its changes, has developed hand in hand with its growing separation from human-centred experience. When we look for the most accurate predictions of the way the world works, they are not to be found in our attempts to understand the activities of society, fluctuations in financial markets, or vagaries of the weather. Rather, it is in describing the interactions of elementary particles or the motions of distant astronomical objects where accuracies of one part in 10^{16} are to be found.

Some sociologists of science have argued that the human contribution to scientific theories is the dominant factor in their success, not their uncovering of any objective reality. But if the latter were true, we would expect our scientific theories to become less and less successful when applied to the extremes of inner and outer space. We would expect to find them at their weakest when applied to environments that were far removed from immediate human experience or the circumstances out of which natural selection has fashioned our senses and sensibilities over millions of years. Exactly the opposite is found. It is in the description of events outside of the direct realm of human experience where our power to predict and explain is best and it is worst in those areas closest to human intuition and experience, by virtue of their intrinsic complexity. Just because there is an undeniable sociology of science does not mean that science is nothing but its sociology.

The course of scientific progress can be seen as a march towards a conception of reality that is divorced from human bias as much as possible. There are several landmarks on this outward journey from us to ultimate reality. First, Copernicus taught us that we should not expect the world to revolve around us—the structure of the Universe guarantees us no special location in space. Then Darwin taught us that we are not the culmination of any special design, and Lyell discovered that most of Earth's geological history went by, rather eventfully, without us. These insights do not mean that our location in the Universe cannot be special in *some* ways—we could not expect to live in a place where life is impossible, like the centre of a star, for example. But our location must not be special in every way. We know that our location in time is indeed rather special, in a niche of cosmic history about 13.7 billion years since the Universe's expansion began, after the stars first formed but before they die. This is why we should not be surprised to find our Universe to be so big and old.

Deeper still was the insight of Einstein, who showed how to express the laws of Nature so that they look the same to all observers, no matter where they are or how they are moving. Newton's famous laws of motion did not possess this universal expression. They would only be seen to take their simple form by special observers who move in a simple way, without acceleration or rotation. For these special observers, the Universe's laws would appear simpler than they would for others. Such an undemocratic situation was a signal to Einstein that something was wrong in our conception of Nature's laws. And he was right. Now we express the basic laws of Nature in forms that would be found by all observers investigating the Universe, from Vega to Vegas, wherever they are, whenever they look, no matter how they are moving. This is the second step.

The third great step in the divorce of science from human idiosyncrasy occurred when a further ingredient was recognized. Besides the laws of Nature and their outcomes, the structure of the Universe around us is determined by a collection of unchanging qualities that we can encode in a list of numbers that we call the 'constants of Nature'. These qualities include things such as the masses of the smallest subatomic particles, the strengths of the forces of Nature, and the speed of light in a vacuum. They are quantified by ever-more-precise measurement, and in the backs of physics books the world over you will find their latest values listed to large numbers of decimal places. These quantities generally have units—the speed of light is measured in metres per second or furlongs per fortnight—which are often rather anthropocentric: centimetres, feet, and inches are conveniently related to the scale of the human frame. Or, equally, they may be geocentric or heliocentric in origin—days and years are units of time that derive from the time for the Earth to rotate once on its axis and to orbit the Sun. These constants are far from universal. They were defined by properties of pieces of metal or the lengths of standard metres kept in special containers in laboratories on Earth. But gradually, physicists realized that the universal constants of Nature allowed standards of mass, length, and time to be defined that did not depend on particular human-made artefacts. By counting the wavelengths of light emitted by a certain species of atom, or counting its vibrations, or the mass of its nucleus, it is possible to define units of length, time, and mass, which can be communicated through interstellar space to physicists who had never seen Earth or their human counterparts.

This march towards established constants of Nature that were not explicitly anthropocentric, but based on the discovery and definition of universal constants of Nature, can be seen as a super-Copernican step. The fabric of the Universe and the pivotal structure of universal laws were seen to flow from standards and invariants that were truly superhuman and extraterrestrial. The fundamental standard of time in Nature, just 10^{-43} of our seconds and defined by the gravitational, quantum, and relativistic constants of Nature, bore no simple relation to the ages of man and woman; no link to the periods of days, months, and years that defined our calendars; and was too short to allow any possibility of direct measurement.

These steps have depersonalized physics and astronomy in the sense that they attempt to classify and understand the things in the Universe with reference only to principles that hold for any observer anywhere. If we have identified those constants and laws correctly, then they provide us with the only basis we know on which to base a dialogue with extraterrestrial intelligences other than ourselves. They will be the ultimate shared experience for everyone who inhabits our Universe.

Modern cosmology makes one further tantalizing suggestion about the nature of the Universe. Before the inception of Einstein's general theory of relativity, all theories of physics were of a similar sort. They provided mathematical formulae that could be used to predict how things would move or change when they encountered other things. They described the action of forces, such as gravity, magnetism, and motion. But in all cases, these laws described the actions of the forces and motions *in* the Universe and within its prespecified space and time. No motion or force could alter the nature of space or of time. They were fixed: God-given and eternal.

Einstein changed all that. His theory is far more sophisticated. When the particles and their motions are introduced into a world governed by the general theory of relativity, they dictate the very geometry of the space and the flow of time. This curved space and time dictates how matter and energy can move, and its motion in turn tells space and time how to curve. It is this feature that gives Einstein's theory its most remarkable quality. *Every solution of Einstein's equations describes an entire Universe.* Some are very simple—too simple to describe our Universe as a whole, but very useful for describing parts of it; some are more elaborate and provide us with wonderfully accurate descriptions of our entire visible Universe. Others describe universes different from our own and impress on us the remarkable nature of its special properties. We hear a lot about that accurate description of our Universe, of its past and its present, and of what to expect in the far, far future. But it has passed unnoticed how remarkable it is that a mathematical theory, a collection of pen strokes on a piece of paper, can provide a description of an entire Universe. The fact that there can exist a mathematical structure of which our whole Universe is a particular outcome is rather astonishing. There could be no stronger evidence of the inadequacy of materialism and no better argument for the reality of a logic behind the appearances that is larger than visible reality itself. How amazing that the mathematical structure that appears to be something bigger than the astronomical Universe itself is the very means by which we can understand its workings. Superhuman the Universe may be, but the ultimate simplicity of the mathematical reality at its heart is what enables us to understand it and have faith that our understanding can converge on the truth.

5 | The natural history of noise

Music creates order out of chaos; for rhythm imposes unanimity upon the divergent; melody imposes continuity upon the disjointed, and harmony imposes compatibility upon the incongruous.

YEHUDI MENUHIN

The club of queer trades: soundscapes

Music, however, as an extraverbal mode of mental functioning, permits a specific, subtle regression to preverbal, that is, to truly primitive forms of mental experience while at the same time remaining socially and aesthetically acceptable.

Heinz Kohut

There have been cultures without counting, cultures without painting, cultures bereft of the wheel or the written word, but never a culture without music. Music, scented sound, is all around us, between our ears and at our fingertips; moving us from head to toe. Without consciously learning its rules, or divining its deep structure, we can respond to the rhythms of a lullaby, be aroused by a call to arms, or gripped by Beethoven's Fifth Symphony. Age is no barrier. Musical ability among the very young, like mathematical genius, can be alarmingly sophisticated, and quite out of step with other skills. But, whereas nobody finds that doing a little bit of long division aids their concentration on other things, musical accompaniment often enhances our completion of other tasks. One reason for the breadth of music's influence is the vast range of sound levels and frequencies that its patterns span: from simple, repetitive drumming to symphonic works of enormous complexity, in which the mental powers and dexterity of dozens of individuals combine to recreate the patterns encoded in its score.

The oldest known musical instruments have been found in Cro-Magnon settlements in central and north-western Europe. They are decorated flutes made from mammoth bones and simple percussive instruments like castenets, and are between 20 000 and 29 000 years old. Other artefacts found with them indicate that these instruments were used in the performance of a ceremony. All known human cultures have well-developed musical practices.

Upon finding transcultural human activities—like writing, speaking, and counting—that display many common features, it often pays to look for ways in which those activities might have evolved from simpler ones whose persistence is biologically advantageous. If the simple predecessor of today's complex activity endowed its exponents with a clear advantage in life—because it made them safer, healthier, or simply happier—then it is likely to become more prevalent because of its cultural transmission; or, if it derives from some inheritable genetic trait that increases fecundity, by being more likely to survive and be inherited. Ultimately, we seek to identify aspects of the physical world that impress themselves upon the human mind more and more firmly over the generations, because a faithful mental impression of them reduces the risks to life that are created by changes in the environment.

At first sight, it is not easy to see what advantage is conferred by a penchant for Beethoven or the Beatles. What could have been the utility of such an abstract and elaborate form of sound generation and appreciation? There is no simple answer. Our impressions are overlaid by many thousands of years of growing complexity and idiosyncrasy. Nor are such questions confined to the origin of music. We can ask them of all the fine arts. If we could strip away our own cultural embellishments, we might be able to see their beginnings in more prosaic practices, which were advantageous to their practitioners. However, even if they aided survival in the distant past, this does not mean that they need play a similar role now.

Painting appears to be a natural outgrowth of the fallibility of human memory, and the need to communicate. Pictures can convey information about the whereabouts of food, or danger; they allow a family, or a group, to inherit and accumulate experience. This is not to deny that we find other, less familiar imperatives in the minds and hearts of image-makers. In ancient times, there was often no sharp divide between the thing that was artistically represented and the representation itself. Many cultures believed that the fabrication or naming of an image gave them power over it. From such beliefs sprang many traditions and prejudices about naming things and people. One influential culture, that of the early Hebrews, refrained from making any artistic images of living things at all—even though they indulged in music with considerable enthusiasm.

Literature and creative writing also seem to have natural precursors in the craving for social cohesion and well-being that can be met by an oral history, or by the telling of stories in which the hearers appear in a leading role. Such tales help to disarm the unknown; they endow life with meaning; they move back the frontiers of the unknown, and promote the self-confidence that comes when sense is made of the world. Their effectiveness is increased by retelling, and the significance of the things recounted is steadily and surely enhanced as a result.

These activities are advantageous if the information they enshrine about the world is true and useful. But false beliefs can also be helpful, as long as they do not inspire fatal activities; they, too, can encourage social cohesion and shared beliefs. This community spirit produces resilience in the face of outside pressures. The knowledge that heroic deeds are recorded and revered encourages acts of bravery and self-sacrifice that otherwise would run contrary to the individual's sense of self-preservation.

In the plastic arts, like sculpture, it is easy to see a link with the development of advantageous skills. The fashioning of tools, weapons, harpoons, and spearheads was an activity in which the best designs, the most robust materials, and the most economical manufacturing processes were a matter of life or death for the participants. The building of shelters encouraged the exploitation of various materials, from clay to wood, to stone and metals. These materials have a spectrum of textures and properties that require a variety of techniques to be invented, evaluated, and refined. There were other reasons for shaping pieces of the world: the search for personal significance, the celebration of human fertility, and the worship of the overt forces of Nature; all seem to bring about a desire to fabricate images. Idols and deities small enough to fit in your home, or round your neck, abound in primitive cultures the world over—indeed, they persist all over the modern world as well. Again, the fashioning of relics plays a powerful, albeit sometimes irrational, role in binding small communities together in ways that distinguish them from other groups.

Another activity that can be viewed in this pragmatic light is that of dance. Whenever there is a need for frenzied activity or heightened sensibilities—in preparation for war, in celebrations of fertility or of birth, or in mourning death—the rhythmic gyrations of primitive dance bind people together in shared experience. The whole community seems larger than the aggregate of its parts; the individual becomes part of a larger dynamic movement that is bound, by solidarity, to the group. These practices offer advantages that are not available to outsiders. They instil order and mutual reliance; they sweep aside the insecurity and hesitation that introversion engenders; but, above all, they offer plausible initial conditions from which some of the rich diversity of civilization can blossom and grow.

The ubiquity of dance is often linked to attempts to make contact with spiritual powers. Anthropologists have reported that it is common for the spirits to be summoned by the beating of a drum. Accordingly, there is usually a close link between the sound of a drum and the marking of a death. Rhythmic drumming has a powerful effect upon us, and we invariably signal approval or disapproval by clapping our hands. When the drumming is loud, we feel the reverberations as well as hear them. It is easy to believe that these sounds would have been the first that humans would have created artificially. They are simple to produce.

They can be made by the hands alone, or by using sticks and stones. Percussion is a basic phenomenon. It is always present in ancient ceremonies of initiation, or attempts to make contact with other realms. The drumming seems to aid the attainment of ecstatic or trance-like states, and it encourages synchronized collective activities like dancing. But perhaps the inner beat of the human heart is important as well. In energetic activities of any sort, the pounding of the heart would become noticeable. Its drumming would link these exciting activities to their inner being. The sexual impetus provided by these activities would have certainly made them adaptive—and we still find a close link between sexual display and loud, rhythmic music. But rhythmic sound might also aid the learning process. If certain memories can be overlaid by an emotional signature then they will be retained more easily ('the everyone-can-remember-what-they-were-doing-when-they-heard-that-J. F. Kennedy-had-been-shot effect').

It is possible that music was originally a special language for making contact with the celestial realm. Sound always seems to be the medium through which to make contact with the gods. The noise of wind and thunder suggests that the gods speak with dramatic force. Many primitive rituals and ceremonies were conducted after dark, when the ear is relatively more important as an organ of sense. A blind person could participate in ancient ritual; a deaf one could not. (The Latin word *surdus*, meaning deaf or mute, is the kernel of our word 'absurd'.)

Music invites us to explore the antecedents from which its appreciation might evolve or accidently spring. There are plenty of possibilities. The earliest, most spontaneous, of human sounds are the cries of a baby at birth, when hungry, or distressed—sounds that we respond to in circumstances of great intimacy. It has been suggested that these whimperings impressed upon us a sensitivity for particular sounds, and developed into a disposition towards musical sounds. Yet humans of all ages retain an ability to make sounds and emotional cries not dissimilar from an infant's cries for attention, and there is no similarity between those cries and music. We recognize our instinctive reaction to crying as one of irritation, unease, or distress—just the reaction we might have expected this experience to have impressed upon our ancestors—not the response that most forms of music arouse within our minds. Despite this distinction, there is undeniably some prenatal conditioning of the human foetus to the body rhythms of the mother, because they are regular enough to be recognized in the presence of other irregular noises. Moreover, these body rhythms restrict our music in definite ways. The division of melodies into musical phrases tends to produce intervals of time that are similar to the human breathing cycle; an even closer approach to this cycle results if singing or wind instruments are involved in the production of sound. The growth of our bodies results in a slowing of our pulse-rates with age. It is probably no accident that young people feel readily at home with a faster musical beat than they will like in later life. If music arose first as an accompaniment to human dancing, then the pulse of the first music would have been dictated by the frequency with which different rhythmic movements could be made.

Another clue to music's antecedents may lie in its emotive power—a power that grows with repeated exposure. In civilizations ancient and modern, the world over, we find the sound of music whenever there is a need to increase group bonding or inspire acts of courage. It creates an atmosphere within which ideas and signals can make a strong impression upon the mind. But therein lies a paradox. For we find that music calms the overwrought human mind just as effectively as it can rouse it. This dichotomy suggests that we shall not find the source of all musical performance, or musical appreciation, in so specific a function as arousal or pacification. Perhaps, as we discussed when considering our mixed responses to the colour 'red', this ambiguity is in itself the most important primer of our attention. Music is sometimes used by psychoanalysts as a form of therapy for mentally troubled patients. This tradition goes back at least as far as Sigmund Freud who, despite loathing music, regarded it as a vehicle that could release mental tensions and speed the return of the psyche to that blissful equilibrium which, for him, was typified by the intimate union of mother and baby.

Since our present perception of music is shaded by the wide range of media in which it is represented, and the symbolic notations it employs, we must not forget that the first music was what we would now call 'folk' music: music that had not been deliberately composed or written down. It was not made for study or appreciation in the modern fashion: you listened to it only in order to learn how to participate in its performance. Such forms of music played a social role that is now regarded as a minor feature of most musical performance—unless you are a football supporter. This change of role shows what a highly structured form music has become. It has evolved far from its original function. And, in so doing, it has become the most theoretical and formally structured of our major art-forms. While the prospective painter or writer can begin ambitious creative work at once, the aspiring musician must be immersed more deeply in the rules and theory of music before any coherent beginning is possible. But, despite the special discipline required of its composers and performers, music can be appreciated without any schooling at all. More than any of the arts, it offers great rewards for little or no prior investment in knowledge.

A ubiquitous source of sound is the inanimate natural world: the wind, the rush of running water, or the crash of thunder. But do they have anything to do with music? There are certainly plenty of sounds in Nature, but most of them are sounds that impede human communication; they are not templates for human emulation. They are copied only in very specific circumstances—attempting to camouflage your presence while hunting, or hiding from enemies—and these activities can be readily distinguished from music-making. More harmonious sounds can be heard elsewhere in the living world. Mating calls and complex bird-song play a key role in the evolutionary process: sexual availability is signalled, mates are attracted, and territory demarcated.

Bird-song turns out to be quite elaborate. There is a definite pattern of development as the bird matures, culminating in its final song. Superficially, this is not unlike the step-by-step development of language in human infants. Some bird species display just one local song, and all the birds learn it; others exhibit a range of different songs and 'dialects' that are influenced by local environmental conditions. (Whale 'songs' are similar in this respect.) Neurological studies of birds reveal that their singing ability is localized in the left part of their brains; damage to that part of a bird's brain eliminates its ability to sing. The songs of a particular species are not inbuilt, because birds of one species can be taught the songs of another. Domesticated birds can be exposed to human 'songs' and will learn them without any instinctive resistance. Some transcripts of bird-song made by the biologist William Thorpe are shown in Figure 5.1.

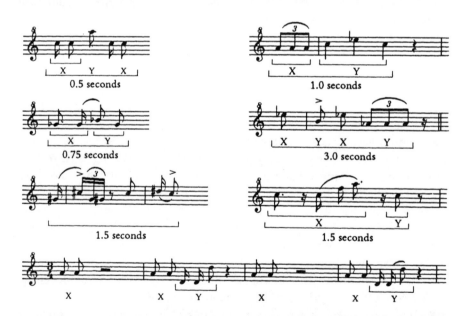

5.1 Seven pieces of bird-song, recorded by William Thorpe. Pairs of African shrikes compulsively sing duets together. Either sex can begin the song, sing it all, or just a part of it. Alternatively, both birds can sing the whole song in time together. These scores display extracts of various durations. The contributions of the two birds to the songs are labelled X and Y.

Charles Darwin favoured trying to explain music by appealing to its possible origin in animal mating calls. Since music has a powerful emotional effect upon us, it may derive from activities associated with attempts to attract mates, with all the heightened emotions and negative feelings of jealousy that go with them. Even now, some singing is associated with feelings of love, especially the sorrow of its loss, or of unrequited love. Darwin believed that music was a primitive precursor of language, whose earliest function was the attraction of mates, but from which sophisticated linguistic abilities subsequently evolved. Mating calls and songs are examples of *sexual* selection, rather than natural selection. Like courtship displays, they play a role in attracting mates, but need not provide information about the genetic attributes of the displayer (although in some cases they may: the male gardener bird (Plate 21) who has constructed the biggest and most elaborate nest is likely to be the fittest and strongest; the amorous toad with the deepest croak will also be the biggest). Sexual selection affects external features of our make-up that influence sexual preferences, and so any art-form that copies or embellishes such fashions derives many of its idiosyncrasies from sexual, rather than natural, selection. Yet, even if an art-form originates in this way, it can subsequently develop by employing representations that are not visually attractive in order to convey a message to the viewer or the listener. In this way, art diverged far from the imperatives of sexual selection. The philosopher Victor Zuckerkandl saw clearly that beauty, while often sufficient for art, is by no means necessary to accomplish its purposes:

Art does not aim at beauty; it *uses* beauty—occasionally; on other occasions it uses ugliness. Art—no less than philosophy or science or religion, or any other of the higher endeavors of the human mind—aims ultimately at knowledge; at truth.

Until late in the eighteenth century, philosophers were much exercised in debate about the extent to which art in general, and music in particular, copies Nature and life. To us, this seems a narrow perspective. For, while there is a plethora of sounds in the natural world, they seem to have little in common with the patterns of tones that we find so enjoyable, and they have not given rise to a specific form of studied listening. Nature almost never gives rise to musical tones. Appreciation of Nature focuses more often upon its serenity than upon the rumble of thunder or the howl of the wind.

Sense and sensibility: a matter of timing

Nothing soothes me more after a long and maddening course of pianoforte recitals than to sit and have my teeth drilled.

George Bernard Shaw

The mind has found ways to make sense of time by linking chains of events into a history. Legends and traditions first played this role, and complemented the mind's ability to make sense of the space around it. The spatial order exhibited in painting or sculpture is heightened when endowed with a temporal aspect. Films are often more appealing than still photographs, and present-day children are virtually addicted to video games. Unchanging images leave viewers to look for themselves. They can look again and again, following first one scanning sequence, then another.* But music imposes its own perceptual order. It has a beginning and an end. A painting does not.

Musical appreciation may be associated with a propensity of the mind to structure time in order to sort and store information. Music-making is more complex, because it requires the coordinated action of different limbs or muscles. Thus, it could be that our liking for music is merely a byproduct of an advantageous adaptation for coordinated actions. What sort of advantage could such an adaptation offer?

'Timing' lies at the heart of all kinds of human activities, from throwing footballs to riding a bicycle. All our complex activities—those that require meticulous coordination of eye, brain, and hand—become acts of exquisite sequential timing when examined in detail. Consider something as 'simple' as crossing the road. We receive visual information and sounds from vehicles moving with respect to us in various directions at unknown speeds. We need to evaluate whether there is an interval between successive vehicles in which we can safely cross the road; we must then move at an appropriate speed to get to the other side—remaining open to the possibility that we shall need to update all the previous information if something unexpected happens. Not only can we do this instantaneously, on roads with curves and gradients, and in conditions of variable visibility, but we can hold a conversation and eat an ice-cream at the same time. The brain has clearly developed an extraordinary facility for sequential and parallel timing of different movements, combining them to produce a single continuous activity like that required to serve a tennis ball. Serious sports competitors will recognize how this timing facility can be improved by careful repetition. A sprinter who has not done any fast running for a while will find, on resuming, that his sprinting action has become rather clumsy and ragged. When the brain sends out signals to the limbs, too many instructions are transmitted

---

* There are particular exceptions, like patterned friezes or op art pictures. In the former, the symmetry is so overwhelming that the brain is entrained by it. In op art, some of the pattern-recognition attributes of the brain are exploited and confused by deliberate ambiguities. For example, we tend to join up the dots, creating imaginary lines between points and their nearest neighbours. But if a pattern is created in which some points have more than one equidistant nearest neighbour, then the eye will flit back and forth along the two possible imaginary lines between the nearest neighbours and the picture can appear to be dynamic (see Figure 2.5).

and rapid forward motion is compromised by all sorts of other unwanted movements: unnecessary muscles are tensed, the head bobs up and down, and the arms flail about in unwanted ways. But, by running many times at close to maximum effort, the nervous system gradually smoothes things out by discarding unproductive movements. In this way, the body can be conditioned to bring into play only the optimal sequence of essential movements. This activity is called 'training'. In fact, in some technical sporting events, it is possible to improve the performance of complicated sequences of movement merely by visualizing the exercise.

Continuity and synchronization are the keys to complex physical performance. Music performance is linked to the development of the brain's facility for coordinating actions that require delicate timing skills. One popular theory of the development of consciousness, espoused by Gerald Edelman, sees the brain as a system undergoing Darwinian evolution by exploring many possible neural interconnections, some of which prove more beneficial than others. These connections are reinforced by use and reuse, at the expense of others. The spectrum of mental activities in which the brain indulges clearly influences its propensity for certain varieties of association. Perhaps the form of timing, association, and temporal organization that music reflects plays an important role in that whole process of neural Darwinism, giving substance to Igor Stravinsky's famous claim that 'Music's exclusive function is to structure the flow of time and keep order in it . . . music is the art of the permutation of time.'

Many animals possess superior coordination and timing: apes perform gymnastics beyond our wildest dreams, and a bear can fish a salmon from a fast-running stream with a regularity that brings tears to an angler's eyes; but neither animal seems to be terribly musical. This suggests that musical intuition is more closely linked to a uniquely human skill, like language—which is also a triumph of coordination between the brain, lungs, chest muscles, larynx, facial muscles, and ears. Just as there appears to be a universal genetic programming of humans that endows linguistic skills, so there might be a universal musical grammar that plays the same role for sound patterns. But, because musical ability is so much more limited in its distribution, it is more credible to believe that it was a by-product of early programming for language than that it is a separate piece of programming.

Before leaving the link between music and temporal ordering, we should note how this association created a deep theological problem for medieval Christian thinkers. It entered their debates about the nature of God and his relation to time and eternity. Music created a dilemma because, if God resides in a transcendent timeless eternity, music cannot form part of the Divine essence; for, without the passing and beating of time, there could be no music. Yet biblical references to heavenly choirs of angels, the prevalence of music in worship, and a belief that we could not be more privileged than God by experiencing music when he does not, led others to conclude that God must share in the temporality needed for musical appreciation.

Incidental music: a harmless byproduct?

Music is essentially useless, as life is; but both have an ideal extension which lends utility to its conditions.

George Santayana

Our dreams provide a window on the mind's attempts to link experiences and events together. If you have ever engaged in any project that requires the cross-referencing of information on many pieces of paper—whether it be completing your tax return or writing an essay—you will appreciate that wonderful cathartic feeling that comes when the project is complete and all the disparate papers can be filed away. Dreams feel like a similar scanning, sorting, and associating process: one that links recent experiences to those of the past. Sometimes, the links take place only at single points, and produce incongruous juxtapositions elsewhere: the mind's attempts to interpolate between single events to create a 'story' often produce bizarre results. Perhaps music affects us so deeply because it resonates with a similar tendency of the subconscious mind to order our auditory experiences. For this reorganization of outside stimuli to take place, it helps to be temporarily insulated from external influences. Music offers this shielding and, thereby, helps the mind to sort information. The feeling of expectation, followed by a resolution of the tensions aroused by a piece of complex music, may be associated with a similar pattern of activity at the neurological level. The experience of music alerts our senses to particular forms of ordered sound. By resonating with the natural book-keeping activities of the brain, music is perceived as relaxing, invigorating, and pleasant.

Many people find it easier to study or carry out linguistic and practical activities against a musical background. It is as if some of the ordering activities of the brain are best kept occupied by processing sound signals to prevent them from interfering with the task to hand. Alternatively, the low level of extra processing required to assimilate an already well-ordered information source, like undemanding types of music, may keep things ticking over in a manner that improves concentration and processing efficiency. Improved performance in IQ tests has been claimed by some (and hotly disputed by others) when the candidates have Mozart playing in the background. And, indeed, Mozart himself is said to have asked his wife to read to him while he was composing, as if to distract the left side of the brain with speech-processing while the right side composed unhindered. In the same vein, it is claimed that Carl Orff would not admit a boy into the Vienna Boys' Choir if he had already learnt to read and write—believing, one supposes, that the opportunity to make the musical-processing side of the brain dominate the language-processing side would then have been lost.

We have already seen that, despite its elevated artistic status, musical ability may not be 'for' anything else at all. It could be an entirely useless elaboration of an ability meant for something else. If the brain likes to sort information by associating common factors, then the emotional significance of a piece of music for the listener may derive largely from a context in which it was once heard. Mendelssohn's Wedding March, the National Anthem, or a well-known advertising jingle are all emotionally impressive because they summon up remembrances of earlier exposures to the same tune and recreate some of the past feelings associated with them. On this view, the form and content of a piece of music is quite irrelevant to the emotional character that we perceive it to have. Rather, that character is entirely determined by the context in which it is heard. This view of music has been dubbed the 'Darling, they're playing our tune' theory. But, it is hard to believe that the meaning of music is entirely determined by context in this way, if only because we can understand something of the structure and 'meaning' of a piece of music without being emotionally moved by it in any way. Moreover, people with similar cultural backgrounds but different personal histories may respond in similar ways when hearing the same piece of music for the first time. Context is clearly important, but it is not invariably all-important.

Another objection to a purely contextual interpretation of musical appreciation is the fact that so much music seems ambiguous, or just downright opaque, to the non-expert: no definite feelings or associations are conjured up at all. On the contextual view, we would be forced to conclude that this music was meaningless for the listener, despite the fact that he might still recognize some of its structural features. Of course, in such circumstances, one is always open to elitist criticisms that one is 'unable to appreciate' the music. One detects a little of this attitude, for instance, in Mendelssohn's comments about the ambiguity of music in his letter to Marc Souchay, written in October 1842, where he claims that

People usually complain that music is so ambiguous; that it is so doubtful what they ought to think when they hear it; whereas everyone understands words. With me it is entirely the converse . . . The thoughts which are expressed to me by a piece of music which I love are not too indefinite to be put in words, but on the contrary too definite . . . And so I find, in every attempt to express such thoughts, that something is right, but at the same time something is unsatisfying in all of them . . .

But it is clear that neither musical appreciation, nor any dexterous facility for musical performance, is shared by people so widely, or at a high level of competence, in the way that linguistic abilities are. In such circumstances it is hard to believe that musical abilities are genetically programmed into the brain in the way that linguistic abilities appear to be. The variations in our ability to produce and respond to music are far too great for musical ability to be an essential evolutionary adaptation. Such a diversity is more likely to arise if musical appreciation is a byproduct of mental abilities that were adaptively evolved primarily for other purposes. Could it be that, unlike language, music is something that our ancestors could live without?

In order to appreciate the spectrum of views about the nature and source of meaning in music, we should explain two conflicting theories that mark out the two most extreme opinions. The first of these is a formal version of the contextual ideas that we introduced just now. Bearing the title *referentialism*, it maintains that the true meaning of music is to be found only outside music. It lies neither in its patterns of sound, nor in its relationships to some absolute aesthetic reality; rather, its meaning is to be found solely in the emotions, ideas, and events to which it refers. Thus the role of music is to 'refer' to something extra-musical; its worth is the measure of its success in so doing. This viewpoint was the official theory of the arts in the old Marxist-Leninist states of eastern Europe. Music, like the other arts, had a function: the furtherance of the goals of the State and the motivation of people to act for the common good of society. Its value was defined solely by the extent to which it achieved those goals. If the emotion that music produces is derived only from its internal harmonic structure, and displays no outside referents, then it is incestuous and decadent. One can see how such a view leads to the tight control of artistic activity. For there is now a clear definition of 'bad' music: that which gives rise to the 'wrong' emotions, and inspires the 'wrong' actions and loyalties. An extreme supporter of this view was the great Russian novelist Leo Tolstoy, who believed that all art should be judged solely in terms of its non-artistic subject-matter. Thus, for him, the 'best' musical compositions were marches, folk music, and dance accompaniments, which produced healthy solidarity. The worst, not surprisingly, included a good deal of the classical repertoire. Of Beethoven's Ninth Symphony he had an especially low opinion, claiming that

not only do I not see how the feelings transmitted by the work could unite people not specially trained to submit themselves to its complex hypnotism, but I am unable to imagine to myself a crowd of normal people who could understand anything of this long, confused, and artificial production, except short snatches which are lost in a sea of what is incomprehensible. And therefore, whether I like it or not, I am compelled to conclude that this work belongs to the ranks of bad art . . . [as does] . . . almost all chamber and opera music of our times, beginning especially from Beethoven [Schumann, Berlioz, Liszt, Wagner], by its subject matter devoted to the expression of feelings accessible to people who have developed in themselves an unhealthy, nervous irritation evoked by this exclusive, artificial and complex music.

This philosophy presents music as a language whose sounds and symbols codify emotions about the outside world. While Tolstoy's interpretation of music is the most extreme version of referentialism—turning it into a piece of State propaganda, as much as a theory of aesthetics—more moderate versions of it are prevalent. They focus upon immediate emotional responses to music, rather than upon the larger social reactions. Some musicologists, such as Deryck Cooke, have attempted to formalize this correspondence by identifying particular intervals and patterns of notes that invariably produce specific emotional responses. Thus, Cooke would claim that the minor second induces feelings of spiritless anguish in a context of finality; joy springs from hearing the major third; whereas the sound of the minor third signals stoic acceptance and impending tragedy. On this view, music would be unnecessary if the composer could transmit his feelings to the audience directly in some other, more efficient manner. But, by converting emotion into patterns of sounds, the composer ensures that many people, even those living after him, will experience those self-same emotions. The American author Diane Ackerman explains how some authors use the contextual associations of music deliberately to create a particular working atmosphere:

Some writers become obsessed with cheap and tawdry country-and-western songs, others with some special prelude or tone poem. I think the music they choose creates a mental frame around the essence of the book. Every time the music plays, it re-creates the emotional terrain the writer knows the book to live in. Acting as a mnemonic of sorts, it guides a fetishistic listener to the identical state of alert calm, which a brain scan would probably show.

Although the full-blown referential view of music sounds extreme, it lurks, thinly disguised, behind the widespread idea that there is a 'message' in a piece of music, or that it is 'good' for children to learn a musical instrument because music contains non-musical values that are beneficial or instructive.

The polar opposite to referentialism is *absolutism*. This looks for the value and meaning of music in those intrinsic qualities that make it an artistic creation, rather than in its context. The patterns of sound make music meaningful; only by attending to those sounds, and excluding all outside allusions, can its unalloyed significance be fathomed. For the absolutist, what truly matters is all that the referentialist counts as nought—and vice versa. The most extreme version of this absolutism is *musical formalism*. It sees a meaning in music that is not to be found in any other human experience. Music is not a representation of something else, and its appreciation should be seen as a higher form of abstract intellectual experience, not unlike 'The Glass Bead Game' of Hermann Hesse's novel of the same name, in which an intellectual elite struggle to produce mental symphonies of meaning that combine musical, mathematical, and intellectual concepts in a sweeping abstract game whose form, while never fully revealed to the reader, creates an exquisite variety of abstract musical chess with its own narrative structure. Presumably it was no accident that the master of this game—the Magister Ludis—was inducted into the religious order of the Glass Bead Game by the Music Master, who discerned the potential displayed by his early musical prowess.

The formalist does not deny that music has external motivations and resonances with extra-musical emotions; it just finds them irrelevant. This view is not unique to music. The philosopher Roger Fry draws on it when considering the content of painting:

no one who has a real understanding of the art of painting attaches any importance to what we call the subject of a picture—what is represented . . . [because] . . . all depends on *how* it is represented, *nothing* on what. Rembrandt expressed his profoundest feelings just as well when he painted a carcass hanging up in a butcher's shop as when he painted the Crucifixion or his mistress.

In music, the formalist view of aesthetics is particularly attractive, because the listener is undistracted by the peripheral machinery of representation, or by choice of scanning order. True musical appreciation must be unencumbered by human emotions and aspirations, because the formalist maintains that these emotions cannot be represented by music. There is no affinity between what we call 'beauty' in the natural world, and musical beauty. This gnostic view of the internal structure of music leads to a rather elitist form of musical appreciation. True musical appreciation is the enjoyment of the pure aesthetic forms inherent in music by those listeners who are sensitive to them. Most listeners are incapable of responding in this way, and so satisfy themselves by dwelling upon the inferior contextual allusions of the music—that is, to all aspects that the referentialist holds dear. The farther from life and human experience the music resides, so the greater is its formal beauty taken to be.

Both of these extreme philosophies of music seem unsatisfactory because of their total exclusion of what the other offers. An alternative philosophy, *expressionism*, takes a middle way, without attempting to be a compromise position. It sees, in music, an aesthetic quality similar to that found in other aspects of human experience. Value in music and experience is to be found in the relationships between them. In this way, the expressionist attempts to come to terms with the puzzle of how a musical work can be meaningful as music and as a human emotional experience. Emotions are aroused when some potential response is prevented or inhibited. Within particular musical traditions, some chords are always followed by others, and the great composers are those who are most skilled at heightening emotional expectations, by postponing and elaborating their resolution. To Western ears this type of postponed denouement is taken to extremes in Indian classical music, where a dissonance will be embroidered and elaborated at enormous length before being finally resolved.

The glass bead game: the music of the spheres

> I consider that music is, by its very nature, powerless to *express* anything at all, whether a feeling, an attitude of mind, a psychological mood, a phenomenon of nature, . . . if, as is nearly always the case, music appears to express something, this is only an illusion, and not a reality.
>
> Igor Stravinsky

The most fervent synthesizers of knowledge were the early Pythagoreans. In the fifth century BC they were among the first to contemplate what we would call 'pure mathematics': mathematical relationships for their own sakes, rather than for some practical purpose. But despite their predilection for arithmetic and geometry, they differed from modern mathematicians in seeing the significance of mathematics to lie in the numbers and geometrical shapes themselves, rather than in the relationships between them. Pythagoras was attracted to the study of musical harmony because it enshrined numerical relationships that could be found elsewhere in the Universe. Thus, deep connections between otherwise unconnected parts of reality seemed to be emerging. His legendary discovery of the simple arithmetical ratios between harmonic intervals persuaded him that there must be an intimate link between mathematics and music—that music was the sound of mathematics, no less.

An ancient account exists, possibly apocryphal, of how Pythagoras discovered the link between number and harmony; Iamblichus tells that

Once [Pythagoras] was intently considering music, and reasoning with himself whether it would be possible to devise some instrumental assistance to the sense of hearing, so as to systematize it, as sight is made precise by the compass, rule and surveyor's instrument, or touch is made reckonable by balance and measures—so thinking of these things Pythagoras happened to pass by a brazier's shop, where he heard the hammers beating out a piece of iron on an anvil, producing sounds that harmonized, except one.

Impressed by the harmonious scale of sounds from the beating hammers, Pythagoras went into the iron-worker's shop to discover how this untutored hammering could produce harmoniously related sounds. He found that the musical intervals they produced were in proportion to the ratio of the weights of the hammers. He went home to experiment further, by hanging different weights on strings of adjustable lengths, and plucked the strings so as to produce different sounds. He discovered that the most appealing sequences of musical tones were linked by simple arithmetical ratios of whole numbers that he and his followers revered. Thus was the numerological link between number and music forged in the brazier's shop.

Pythagoras is credited with the discovery that notes in harmonious relation can be produced by plucking strings whose lengths are in particular ratios to each other. The shorter the string, the higher the note. Halving the length of a vibrating string produces a note that is higher by one octave; doubling it produces a note that is an octave lower. The ear seems to like the combinations of notes produced by strings whose lengths are in the ratios 1:1, 1:2, 2:3 (the 'perfect fifth'), or 3:4. Pick a ratio like 7:11 and the result is noticeably discordant. Pythagoras could determine the ratios of the lengths of string required to produce combinations that were agreeable to the ear. In this way, the Pythagoreans' religious reverence for numbers was overstimulated, and the belief that numbers each possess an encrypted meaning remained strongly associated with the study of musical harmony for nearly two thousand years. The Pythagorean union of mathematics and music was first taken up by Plato and, together with the mathematical description of the motions of the heavenly bodies, became the basis for a cosmological picture in which the harmonies of music, mathematics, and celestial movement were inextricably linked. This chain of thinking was one of the most extreme forms of reductionism ever conceived. Since musical tones and celestial motions both displayed mathematical relationships, it was believed that they must be equivalent to one other at some level. From this it was argued that each of the moving celestial bodies must produce musical tones that will depend upon the distance of the body from the Earth, and upon its speed. Moreover, these tones combine to create a celestial harmony: 'the music of the spheres' (Figure 5.2). Aristotle describes the reasons for this view in his work *De Caelo* ('The Heavens'):*

the motion of bodies of that [astronomical] size must produce a noise, since on our Earth the motion of bodies far inferior in size and speed of movement has that effect. Also, when the sun and the moon, they say, and all the stars, so great in number and in size, are moving with so rapid a motion, how should they not produce a sound immensely great? Starting from this argument, and from the observation that their speeds, as measured by their distances, are in the same ratio as musical concordances, they assert that the sound given the circular movement of the stars is as harmony.

In the first and second centuries AD, there was serious scholarly debate as to why we cannot hear this celestial music. Some argued that it was outside the range of human hearing, others that its ubiquity meant that we were oblivious to it, and

---

* *The Works of Aristotle*, Vol. 2, trans. J. L. Stocks (Oxford University Press, 1930).

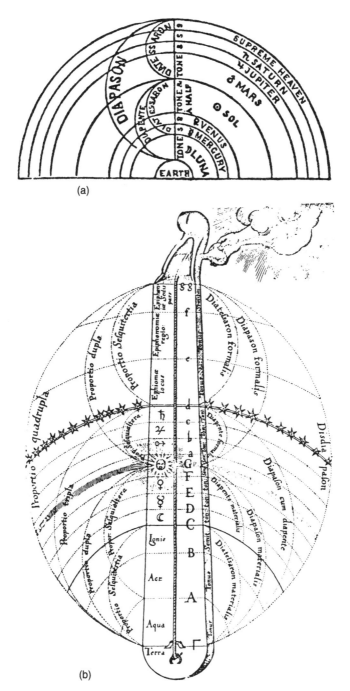

5.2 (a) A Pythagorean division of the celestial sphere into musical intervals. (b) A medieval elaboration of the Pythagorean ideal of harmony between humanity and the environment as displayed in Robert Fludd's 'The Tuning of the World', from *Ultriusque Cosmi Historia*.

heard only changes in sound relative to it. Others maintained that its loudness has made us deaf to it. None of these theories seems to have won widespread acceptance.

This ancient belief in a cosmos composed of spheres, producing music as angels guided them through the heavens, was still flourishing in Elizabethan times. It is most eloquently espoused by Shakespeare, in *The Merchant of Venice*. Approaching Portia's house, Lorenzo describes the celestial harmony to Launcelot; our deafness to it is a consequence of our mortality:

> How sweet the moonlight sleeps upon this bank!
> Here will we sit and let the sounds of music
> Creep in our ears. Soft stillness and the night
> Become the touches of sweet harmony.
> Sit, Jessica. Look, how the floor of heaven
> Is thick inlaid with patens of bright gold.
> There's not the smallest orb which thou behold'st
> But in his motion like an angel sings,
> Still choiring to the young-eyed cherubins.
> Such harmony is in immortal souls,
> But whilst this muddy vesture of decay
> Doth grossly close it in, we cannot hear it.

There was a good deal more to Pythagorean musical theory than celestial harmony. Besides the music of the celestial spheres (*musica mundana*), two other varieties of music were distinguished: the sound of instruments, like flutes and harps (*musica instrumentalis*), and the continuous unheard music that emanated from the human body (*musica humana*), which arises from a resonance between the body and the soul. The important assumption behind these distinctions, which was taken up by Plato, and then influenced Western philosophy for so long, is that the celestial music exists and has its properties quite independently of the human listener. In Plato's mind, what we hear of musical harmony is a pale reflection of a deeper perfection in the world of number, which displays itself in the planetary motions. We appreciate it only because the rhythms of our body and soul are preformed to resonate with the harmony in the celestial realm. It was this transcendental philosophy of music that Plato reinforced by his wider belief that the world of appearance is a shadow of another perfect world filled with the ideal forms of the things around us. Ultimately, Platonic philosophy is the source of the absolutist philosophy of music, which we discussed earlier.

In the medieval world, the status of music is revealed by its position within the Quadrivium—the fourfold curriculum—alongside arithmetic, geometry, and astronomy. Medieval students of music regarded themselves as scientists, and the relation of music to mathematics and astronomy was regarded as the most important aspect of music. They believed all forms of harmony to derive from a common source. Before Boethius' studies in the ninth century, the idea of musical harmony was not considered independently of wider matters of celestial or ethical harmony. A great change in the view of music could occur only in a new climate that relinquished its total reverence for the authorities of the past and sought to answer questions about things by looking at them, or listening to them, rather than merely reading about them.

In early medieval times, the performance of music was a mundane and secondary business, irrelevant to its true meaning and quality. We are so accustomed to thinking of music as a performing art that it is hard to appreciate that a strong interest in musical performance did not arise until the Renaissance. Another aspect of musical performance that we now take for granted is its mixing of different melodies: that is, *polyphony*. Polyphony—the combination of two or more strands in a musical texture—began with the addition of one or more parts to a plainsong melody. Singing in parallel fifths dates from the eighth century, but independent vocal parts did not appear until the eleventh century. These developments provided the basis of what was eventually to become the elaborate harmonic structure of later music.

The simultaneous sounding of different notes is a strange phenomenon. Mix colours or textures and they lose their individuality by blending; but combinations of musical tones combine without losing their identities. To those engaged in the metaphysical study of music, this must have seemed a deep mystery; yet the huge time it took for polyphonal music to emerge suggests that there was some natural antipathy or ideological barrier to it. Through this development, Western music parted company with other traditions, and evolved comparatively rapidly into structures of vast complexity. Curiously, the thousand years that it took for music to reach the pinnacle of classical complexity that so many people still enjoy today saw a parallel development of its ancient bedfellow—mathematics—to undreamt-of levels of abstract sophistication, which far outstripped contemporary practical application.

Later, the developing complexity of classical polyphonic composition brought with it a human dimension to the previously impersonal and transcendental realm of music. In compositions like those of Beethoven and his gifted contemporaries, we see the expression of the composer's personality in his music. Whereas the quest for the true meaning of music had once looked to transcendental realities in the heavens for ultimate satisfaction, its truths could now be found by introspection and psychology. Music would tell of the inner struggle of its creator, or resonate with the emotions of the listener: amplifying, modifying, or pacifying them in ways that were seen as deriving from the music, not merely arising in response to it. In this way, passionate listeners claim to find a deep meaning in music that transcends all other art-forms. Such was the confidence that humanity had in its own achievements as this new music emerged that, instead of downgrading its status from the music of the spheres to the muse of Mankind, its new emphasis served primarily to upgrade Man's estimation of Man. And thus, as the classical symphony became grander and more elaborate in structure, so its focus and interpretation became more personal and more closely associated with the character of its composer. And with the movement of music away from esoteric notions of celestial harmony, towards personal meaning, its popularity grew far and wide. Great concert halls were needed to accommodate the listeners, and music played a central role in public life all over Europe. But with these institutions, and the strata of society that frequented them, grew up an elitism about music. Much musical performance was exclusive: it was expensive to attend concerts, and in order to appreciate what was performed it was necessary to possess a sensitivity for and appreciation of the social setting of musical performance. What happened in the nineteenth century was a strange turning of things upon their heads. Music was no longer defined or interpreted by its correspondence with perfect geometrical patterns, whether in the sky or on paper. The notes, and even the performers, had become secondary to the effect that the music had upon the hearer. An anti-Copernican revolution had occurred, which placed the human soul and spirit at the fulcrum of interpretation. But this did not last long. With the coming of psychologists like Freud, the status of human responses to something as subjective and overlain with other emotional glosses as music was downgraded into just another form of emotional release from psychological tensions.

By the early years of the twentieth century the possibilities of Western tonal harmony had been thoroughly explored by a dazzling array of gifted composers. It was time for a counter-cultural reaction. It arrived, in 1907, with the first performances of works by Arnold Schoenberg that in their extreme chromaticism stretched the key system to its limit (if not beyond). Later, in the 1920s, Schoenberg was to develop the serial system of composition using twelve tones with which his name is associated.*

Vehement protests arose when they were first performed. The performance of such deliberately atonal music served to accelerate the perception of contemporary music as an obscure, highbrow activity—for initiates only. Gradually, this emphasis, and the strong focus upon the personality of the artist as the primary factor in his work has had a strong negative effect upon the status of music—an effect that can be detected elsewhere in the creative arts as well. For, when personality is all, any period during which eccentric or powerful personalities

---

* In a curious deference to numerological superstition, of which Pythagoras himself would have been proud, Schoenberg used the spelling 'Aron' instead of 'Aaron' in the title of one of his operas, *Moses and Aron* so that it would have twelve, rather than (unlucky) thirteen characters. Ironically, he died on Friday, 13 July 1951.

are absent can be interpreted as an era of blandness in the art-form itself. Classical music no longer plays a central role in our culture. It is not headline news in any sense. It is too far divorced from the centre of gravity of things. The most newsworthy of the arts—popular music—plays a central role in youth culture, but it might be argued that, to a considerable extent, it has also reached that position for reasons that have little to do with its musical content. Again, the focus has always been primarily upon the performers as personalities, or cult figures, rather than as musicians. Their music has served as a rallying-call to counter-cultural movements reacting against established norms of behaviour as a whole, not simply against its musical tastes. The modern era has, however, seen the emergence of a new musical phenomenon: that of the solitary listener. With the availability of music on the radio and the gramophone it became possible to be a private listener. This has countered the elitism of the nineteenth century, and has promoted the study and analysis of music for reasons other than entertainment. It has also enabled far more diversity to develop in musical style. Unusual forms of music, of only minority interest, can be performed and heard without the expense of hiring vast concert halls in which to hold public performances. Ironically, many modern pop works are a profound disappointment in live performance, because of the enormous reliance upon synthesized sound and multi-tracking that studio production can easily produce, but which live performers, in real time, often cannot.

Player piano: hearing by numbers

Music and science were [once] . . . identified so profoundly that anyone who suggested that there was any essential difference between them would have been considered an ignoramus [but now] . . . someone proposing that they have anything in common runs the risk of being labelled a philistine by one group and a dilettante by the other—and, most damning of all, a popularizer by both.

Jamie James

There has long been a suspicion that there exists some deep connection between mathematics and music. Pythagoras started it, and once this genie was out of the bottle it was terribly hard to get it back in again. Thousands of years later, the deep structure in Bach's music inspired Leibniz to claim that 'music is the hidden arithmetical exercise of a soul unconscious that it is calculating'. The origin and development of this idea has shaped attitudes to music over the past two thousand years, and has been discarded as a central paradigm only in the last three hundred years. Looking at music today, there is a superficial similarity between mathematics and music because both make use of symbolic notations (Figure 5.3).

But differences abound: mathematics has a logical inevitability that music lacks; clearer still is the division between the skill of making music, either by composing or performing, and the pleasure that comes from listening. There is no similar division in mathematics. It is not a spectator sport. Only practitioners of mathematical logic enjoy reading or hearing about it. Moreover, mathematical proofs give a somewhat misleading picture of what mathematicians actually do, and how they think. There is a real divide between the creative work of mathematicians and the formal presentation of their results. The differing reactions of an 'audience' to mathematics, as opposed to music, highlights the ability of music to arouse mass emotion and action—an ability that mathematics entirely lacks. This suggests that music is linked to more primitive instinctive responses to the world than is counting.

The multicultural profile of musical performance and appreciation is a striking feature of human civilizations the world over. This universality is shared by a human propensity for language and for counting. Although there are superficial similarities between these human abilities, they impress us more by their differences. Musical tones certainly sound different from words; and the processing of musical tones by the brain differs from that of language. Our reception of tones is interfered with by introducing further tones, but not by adding verbal information in the form of words or numbers. These disparities are displayed at a neurological level by what we know of the geography of the brain. In

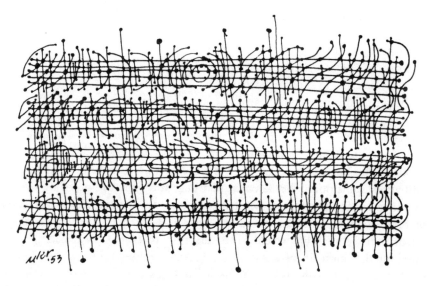

5.3 *Alien Musical Scores*; a drawing by Robert Mueller.

right-handed individuals, linguistic abilities are almost entirely controlled by the left hemisphere of the brain, whereas musical sensitivities are largely governed by the right hemisphere. Accordingly, serious damage to the left side of the brain is generally catastrophic for speech, but leaves musical abilities undisturbed. Conversely, damage to the frontal and temporal lobes of the right hemisphere, or a disease of this side of the brain, is disastrous for musical enjoyment: it reduces our ability to discriminate between sounds, and our appreciation of nuances in pitch. This asymmetry between the two hemispheres is displayed by our hearing as well; sound received by the right ear is processed in the left hemisphere, that by the left ear in the right hemisphere. Hence, we tend to process language more effectively when it is heard with the left ear, and musical sounds entering the right ear are remembered better than those entering the left ear. However, when individuals with considerable musical training are tested in the same way, these differences are considerably reduced. Musical training presumably enhances the potential for analysis of musical structure by means whose provenance lies within the brain's left hemisphere. This is not entirely surprising. We would expect that someone schooled in the mathematical aspects of musical structure would activate some of the mathematical processing networks within the brain when listening to music. In general, if there is some contextual association with an item of mathematics or music, then the language-specific thought-processes that deal with it should be awakened by contemplation of it.

Despite these neurological trends, there are many peculiarities and exceptions, which reflect the diversity of human musical ability. Aspects of musical ability that are strongly affiliated to skills of performance suffer if the parts of the brain governing the associated motor skills are injured. Also, besides our own instrumental performances, music is fired at us from a wide variety of sources—from vocalists, rock groups, orchestras, birds, records, and also as a background to film and dance. The association between music and the 'something else' that goes with it, especially in situations where that association heightens the emotions, is likely to produce very complicated mental responses. By contrast, our exposure to the sound of language is relatively uniform—even recordings of speech sound the same as live voices—and the average person's exposure to mathematics is even less stimulating. This uniformity makes linguistics a far more focused mental ability than musical appreciation.

If we examine our ability to count and calculate from a neurological perspective, then we find there are individuals who lose their linguistic abilities because of brain damage, but can still count. Some of the key mental circuitry for calculation seems to be present in the right hemisphere of the brain, although many of the quasi-linguistic aspects of reading and describing mathematical symbols are dealt with, like language, by the left-hand side of the brain. Some areas of the left side of the brain, which play an important role in spatial orientation, may also be important for number sense and the sort of geometrical intuition that mathematicians prize. Although our simple counting ability may have its origins in the right hemisphere, abstract mathematical reasoning seems to reside in the left hemisphere. This leaves the right hemisphere to control more synthetic and holistic operations, especially those involving images and metaphorical description, together with the processing of music.

The relationships between music, pattern, and language invite us to map out some speculative scenarios for their historical development. Six clear options suggest themselves. In the first, there exists some form of common ancestral mental function among humanity's forebears which split into separate threads—one of music, the other of language—while retaining some residual traces of the link between the two which is manifested in activities like singing. In the second possible scenario, music is assumed to be primary, with language developing from it—perhaps stimulated by physiological or neurological evolution. In the third option, language is primary, and music subsequently evolves from it as a separate activity—for example, because of the development of singing as a means of transmitting sounds over long distances. Fourth, language could be a strand of human activity and culture that developed in parallel to a more basic facility for pattern-recognition. At first, spatial pattern-recognition became well developed and spawned byproducts, like art and image-making; then, temporal pattern-recognition became acute, and diversified into musical rhythm. In this scenario, music develops after other artistic practices. Fifth, there might be a primary facility for pattern-recognition, from which language-making split off. Subsequently, temporal pattern-recognition developed, and spawned a further cultural offshoot in music, while the spatial pattern-recognition thread gave rise to art as a cultural manifestation. Sixth, a primary pattern-recognition ability might gradually have diversified into a sequence of more specialized abilities: first, recognition of spatial pattern, then of temporal sequences, then of language and numerical sequences.

At present, most linguists seem to believe that language is a specific human ability, rather than merely another byproduct of the brain's general pattern-recognition and learning abilities. Impressive evidence can be marshalled to exhibit features shared by disparate human languages, which witness to a universal 'grammar' that is hard-wired into the structure of the brain. This makes sense of the observation that children do not really seem to learn language to an extent that is commensurate with their skill in using it. As we described in Chapter 2, linguistic abilities just seem to be programmed to switch on at particular times in early development. Language is thus seen as a natural instinct rather than a learned behaviour: it is primarily a product of Nature rather than of nurture. We could ask whether the same attribution could be given for either mathematical or musical ability. This notion is much harder to sustain. Musical ability is not shared at the same level of competence, or as ubiquitously, as linguistic ability. One of the most striking things about linguistic ability is how sophisticated and uniform it is compared with all other skills. There are hordes of healthy individuals who cannot add up, or who care little for music of any sort, but none who cannot speak a language. If one examines the languages of traditional peoples, who often had no mathematical systems beyond counting up to two, five, or ten, their language is similar at root to our own and in no sense primitive when gauged against the vocabulary required by their lifestyles. A study of the origins of counting in ancient peoples reveals a common pattern of simple counting systems. One could argue that the systems of number words they employ are primarily linguistic, rather than 'mathematical', in character. In order to graduate to deep and difficult mathematics—rather than merely using symbols as a shorthand for words describing quantities—sophisticated notational concepts are required, and these were introduced by only a few advanced cultures. One of those crucial steps is the invention of a 'place-value' notation for representing numbers, whereby the relative position of a symbol conveys information about the quantity that it represents. Thus, for us, the expression '341' means three hundreds plus four tens plus one unit. This powerful idea, together with the idea of a zero symbol, '0', which it requires, was devised by only three advanced cultures: the Sumerians and Babylonians, the Mayans, and the Indians. Our propensity for this positional notation may be linked to the syntactic programming that our minds possess for natural language. Unfortunately, no detailed studies have yet been made into the links between the linguistic traits in traditional cultures and the structure of their counting systems. Until a careful study of this interplay between the linguistic use of number words is made, it is difficult to determine whether human propensities for counting are really separate from those for language in ancient cultures.

If we compare music with mathematics, then it is clear that music displays a greater cultural diversity than mathematics. This is not surprising if we regard music as a human invention and elaboration, because mathematics appears to offer far more than that. Music does not help us understand the workings of the physical world: mathematics does. Mathematics exhibits a multitude of features that point to some of our mathematical knowledge being the fruit of discovery, rather than merely a byproduct of abilities evolved for other purposes. Mathematics affects us quite differently from music or language. Language is totally flexible; it can affect us emotionally or logically. Music primarily influences our emotions, in ways that mathematics cannot. Yet, each of these three activities is tied in some way to the limits of our physiology. Sophisticated human language was possible only because of the evolution of the special structure of the human larynx that other mammals do not possess. Without this purely anatomical development, no amount of special neural programming for linguistic ability would have been able to help us. In the case of mathematics, we can see how our ten fingers (and in some cases, our ten toes as well) determined the form of many of the counting systems that were first developed. However, the decimal (base-10) system that we have adopted is not the optimal one for all purposes—as the use of binary arithmetic in computer languages shows. Mathematics could have developed quite satisfactorily if the counting-base had been chosen differently (say base-12) in the most influential Indo-European cultures. Similarly, we shall see that appealing music is significantly limited in range, and form, by the sensitivities of the ear, and by the frequency-analysing abilities of the brain. If we wanted to make ourselves understood to extraterrestrials, then we might hope to do so by using mathematics. In order to rely upon our languages, it would have to be the case that our basic grammar, operating at the neurological level in our brain's language-processing software, was the only program capable of performing such linguistic tricks. In the unlikely event that this were so, intelligent extraterrestrials would share the root structure of human mental grammars. Even so, we would have to do a vast amount of analysis and decoding to unravel a message in their language; whereas, a description of a shared physical system—like an atom or a light ray—would allow the superficial differences in mathematical symbolism to be removed much more easily. By contrast, music probably would not help us understand each other or communicate directly at all. Rather, it would reveal important things about the nature and physiology of its generators and appreciators. It is reasonable to speculate that extraterrestrials would possess 'artistic' byproducts of their evolutionary adaptations, but there is no reason why they should be primarily associated with sound signals, as opposed to light signals, motor functions, or even taste buds. The music that we know is a very specialized human byproduct, which is appreciated because of the brain's special adaptations for other aspects of the world, and the need to predict and anticipate the changes that can occur in our environment. Whereas linguistic ability appears to be a necessary consequence of our humanity, music does not seem to display quite the same inevitability or sophistication, and mathematical ability seems to be neither necessary, nor evident, to any significant extent in most humans.

One might be tempted to think that if enjoyable music could be reduced to mathematical patterns of a definite variety then the puzzle of what music 'is' would somehow be solved. Unfortunately, things are never so simple. For it is a well-kept secret of mathematicians that even they do not know what mathematics is. Four philosophies of mathematics are current among mathematicians, philosophers, and users of mathematics. I have argued elsewhere* that the remarkable applicability of mathematics to the structure of the physical world, and the laws

---

* See John D. Barrow, *Pi in the Sky: Counting, Thinking, and Being* (Clarendon Press, 1992).

that govern it, should be taken as the most important datum in deciding between them. To appreciate the depth of the gulf between mathematics and music, despite their superficial similarities and ancient traditions, we need to look more closely at the extraordinary utility of mathematics and how it is best interpreted.

Scientists believe so deeply in the mathematical structure of Nature that it has become an unquestioned article of faith that mathematics is both necessary and sufficient to describe everything from the inner space of elementary particles to the outer space of distant stars and galaxies—even the Universe itself. Yet, why does the symbolic language of mathematics have anything to do with the falling apples, splitting atoms, exploding stars, or fluctuating stock markets? Why does reality march to a mathematical tune? The answers depend crucially upon what we think mathematics is.

At the beginning of the 20th century, mathematicians faced some bewildering problems that rocked their confidence. Bertrand Russell proposed logical para-doxes, like that of the barber,* which threatened to undermine the entire edifice of logic and mathematics. For who could foresee where the next paradox might surface? In the face of such dilemmas, David Hilbert, the foremost mathemat-ician of the day, proposed that we should cease worrying about the *meaning* of mathematics altogether. Instead, we should simply *define* mathematics to be no more, and no less, than the entire collection of formulae that can be deduced from a collection of consistent initial axioms by manipulating the symbols involved according to specified rules. He believed that this procedure could not create paradoxes if it was accurately executed. The vast embroidery of logical connections that results from the manipulation of all the compatible groups of starting axioms in accord with all the possible collections of rules is all that mathematics 'is'. This viewpoint is called *mathematical formalism*. Like musical formalism, it eschews any search for the meaning of a pattern of symbols in a context that is external to their representation. The formalist would no more offer an explanation for the mathematical character of physics than would the musical formalist try to explain why Wagner can be depressing.

Hilbert thought that this strategy would rid mathematics of all its problems—by definition. Given any mathematical statement, we could determine whether it was a valid deduction from a consistent set of starting assumptions by working through the sequence of logical connections. Hilbert and his disciples set to work, confident that they could parcel up all known mathematics with their rules and axioms, leaving Russell's paradoxical chimeras safely beyond its pale. Unfortunately, and totally unexpectedly, their enterprise collapsed rather sud-denly. In 1931, Kurt Gödel, then an unknown young mathematician at the

---

* A barber shaves all those individuals who do not shave themselves. Who shaves the barber?

University of Vienna, showed Hilbert's goal to be unattainable in any mathematical system large enough to include ordinary arithmetic. Whatever set of starting axioms one chooses, whatever set of consistent rules one adopts to manipulate the mathematical symbols involved, there must always exist some statement, framed in the language of those symbols, the truth or falsity of which cannot be decided using those axioms and rules. Worse still, there is no way of ever telling whether the starting axioms are logically consistent or not. Surprisingly, mathematical truth is something larger than axioms and rules. Try solving the problem by adding a new rule, or a new axiom, and you merely create new undecidable statements. If you want to understand logical truth, you have to venture outside mathematics. If a 'religion' is defined to be a system of ideas that contains unprovable statements, then Gödel showed us that mathematics is not only a religion, it is the only religion that can prove itself to be one.

A more interesting close cousin of formalism is *structuralism*. This is the philosophy of mathematics that sees it as the collection of all possible patterns. Some of these patterns are instantiated in physical objects—clouds, wallpaper designs, or the shapes of galaxies; others are in sequences of operations; while others are present in mental operations or properties of those quantities that we call numbers. This view of mathematics runs the risk of being over-inclusive because it includes as mathematics all manner of pattern-making activity—marking lines on the highway, hairdressing, painting, or the assembly-line workers at a car factory—which all mathematicians might not regard as mathematics. But it is a small price to pay for a picture of mathematics that rings true and provides a simple answer to the mystery of why mathematics works so well in describing the world. In order for any thinking beings to exist in the universe there would have to exist some order in it somewhere. What we call mathematics is just the study of that order and the patterns that we are able to generate from it. The mystery of the effectiveness of mathematics is slightly shifted in focus: the real mystery is now seen to be, not that mathematics works, but that such simple mathematics is so powerful and far-reaching in what it can tell us about the world. Structuralism is appealingly simple. It is a useful way to convey what mathematics is about to non-mathematicians. If we try to apply it to other more specialized subjects then its over-inclusive quality becomes a bigger problem. As a philosophy of music it would end up defining music to be collection of all patterns made with sound. The all-inclusiveness draws in explosions, canteens of cutlery falling on the floor, as well as human speech. Modern and experimental music would welcome this inclusive breadth, but the weakness of the definition is evident. What is not music?

A less inflexible picture of mathematics is one that focuses on the fact that it is an open-ended human activity. *Inventionism* is the belief that mathematics is nothing more than what mathematicians do. Mathematical entities, like sets or triangles, would not exist if there were no mathematicians. We invent mathematics; we do not discover it. The inventionist is unimpressed by the utility of mathematics—arguing that the properties well-suited to mathematical description are the only ones that we have been able to uncover. This view of mathematics is common amongst 'consumers' of mathematics, particularly social scientists and economists, because of their preoccupation with artificial social constructions, and their study of problems that are so complicated that many approximations and idealizations are necessary to make them tractable.

The independent discovery of the same mathematical theorems by different mathematicians from totally different economic, cultural, and political backgrounds—often at widely separated times in history—argues against such a simple view. The inventionist could respond by pointing to the universality of human languages. Despite their superficial differences, there is strong evidence that they share a common underlying structure. This 'universal grammar' means that a linguistically sophisticated extraterrestrial visitor to planet Earth would have grounds for concluding that humans spoke a single language, albeit with many regional nuances. One might therefore expect that those aspects of this universal grammar that share features of simple logic, and hence counting, would also make counting appear instinctive. In fact, although simple counting—often to bases other than ten—is fairly universal in ancient and primitive cultures, virtually none of them went on to carry out mathematical operations more sophisticated than counting. This suggests that these higher mathematical operations are not genetically programmed into the human brain—and what possible evolutionary reason could there be for lavishing valuable resources upon such a luxury? They are more likely to be byproducts of multi-purpose pattern-recognition abilities. But simple counting, because it is so closely allied to linguistic operations and the logic of the brain's own programming for language, is effectively programmed in.

Another objection to the inventionist view of mathematics emerges from contemplating the evolutionary origin of our minds. Even if mathematics, in some sense, comes out of our minds, or is imprinted upon them by sensations of the natural phenomena that we witness, what is the source of that mathematical structure? Our minds cannot create it out of nothing. Rather, the mathematical structure of the world is instantiated in the human mind by an evolutionary process that rewards faithful representations of reality with survival, and eliminates unfaithful images of reality because they have low survival value. When traced to its source, inventionism dries up.

An interesting modern variant of inventionism is *social constructivism*, which sees mathematics as something that has emerged out of our collective social interactions. In this respect it is like a national constitution, legal code, or monetary system. These things are real but they are neither physical nor mental. They arise out of the social and cultural interaction of many individuals. They do not reside solely or completely in the mind of any single individual and without the people who gave rise to them these social constructs would not exist. This approach to mathematics appeals to the phenomenon of emergence, which has become a fashionable approach of explaining complexity. It sees mathematics as a complicated social activity which has emerged from simpler, well-defined mental and physical activities and has achieved an astonishingly high degree of consensus among its participants. This view doesn't help especially when it comes to understanding why mathematics works so well as a description of the world or why abstract mathematics so often turns out to be practically useful. Its subscribers would just argue that it enlarges the context in which a solution to those problems can be sought. If we apply this approach to music then it fits rather better. Music can be viewed as an emergent phenomenon that has local roots that form style and pattern. It develops over time. It involves the formation of a collective opinion. But none of these features are unique to music. As with structuralism we have encountered the problem of an over-inclusive definition.

The third option is Platonism. For the mathematical Platonist, the world *is* mathematical in some deep sense. Mathematical concepts exist and are discovered by mathematicians, not invented by them. 'Pi' really is in the sky. Mathematics would exist in the absence of mathematicians, and could be used to communicate with extraterrestrial beings who had developed independently of ourselves. Interestingly, this view seems to be assumed implicitly by all present exponents of the 'Search for Extra-Terrestrial Intelligence' (SETI), who beam into space information that is predicated upon the universality of the concepts underlying human science and mathematics.

Whereas formalism and inventionism are embarrassed by the unreasonable effectiveness of mathematics as a description of Nature, the Platonist makes it the cornerstone of his case. Most scientists and mathematicians carry out their day-to-day work as if Platonic realism were true, even though they might be loath to defend it too strongly when questioned in a pensive mood at the weekend. But mathematical Platonism has its difficulties. It is permeated by vagueness. Where is this other world of mathematical objects which we discover? How do we make contact with it? If mathematical entities really exist beyond the physical world of particulars that we experience, then it would seem that we can make contact with them only by some sort of mystical experience that is more akin to the seance than to science. This means that we cannot treat the acquisition of mathematical knowledge in the way that we treat other forms of knowledge about the physical world. We treat the latter as meaningful knowledge because the objects about which we have knowledge are able to interact with us in some influential way, whereas there seems to be no means by which mathematical entities can affect us, or be influenced by us.

The last response to the ferment of uncertainty about the logical paradoxes that spawned formalism in the early years of this century was *constructivism*. It was a mathematical edition of the doctrine of operationalism. Its starting-point, according to Leopold Kronecker, one of its creators, was the recognition that 'God made the integers, all else is the work of man.' What he meant by this slogan was that we should accept only the simplest possible mathematical notions—that of the whole numbers 1, 2, 3, 4, . . ., and the operation of counting—as a starting-point, and then derive everything else from these intuitively obvious notions step by step. By taking this conservative stance, the constructivists wished to avoid manipulating counter-intuitive entities like infinite sets, about which we could have no concrete experience. As a result, constructivism became known as *intuitionism*, in order to stress its self-stated appeal to the bedrock of human intuition.

For the constructivist, mathematics is the collection of deductions that can be constructed in a finite number of deductive steps from the natural numbers. The 'meaning' of a mathematical formula is simply the finite chain of computations that have been used to construct it. This view may sound harmless enough, but it has dire consequences. It creates a new category of mathematical statement. For the status of any statement can now be threefold: true or false or *undecided*. A statement whose truth cannot be decided in a *finite* number of constructive steps remains in the third, limbo status. Preconstructivist mathematicians, dating back to Euclid, had developed a variety of ways of proving formulae to be true that did not correspond to a finite number of constructive steps. One famous method beloved of the ancient Greeks was the *reductio ad absurdum*. To show something to be true, we assume it to be false, and from that assumption deduce something contradictory (like $2 = 1$). From this we conclude that our original assumption must have been false. This argument is based upon the presumption that a statement is either true or false. But for the constructivist a statement is shown to be true only after explicit demonstration in a finite number of deductive steps.

If we take a long look at constructivism it seems a peculiar doctrine indeed. It is more like a philosophy of a deductive game such as chess, rather than of mathematics. In order for it to work, it has to remove well-established forms of logical argument from the mathematician's armoury. It defines mathematics in an anthropocentric fashion: as the totality of all finite step-by-step deductions from the bedrock of human intuition—the natural numbers. There is no mathematical existence before this process of construction takes place. Besides its anti-Copernican stance, the notion that there exists a universal human 'intuition' for the natural numbers does not have historical support. The constructivist can never tell if my intuition is the same as yours, or whether human intuition has evolved in the past or will evolve further in the future. The mathematics that it creates from human intuition is a time-dependent phenomenon that depends upon the mathematician involved in its construction. Constructive mathematics is close to being a branch of psychology. It raises many problems. Why should we start with the natural numbers? What counts as a possible constructive step? Why are some constructions more useful and applicable to the real world than others? Why can we not have intuitions about infinite collections? How does one explain the utility of non-constructive concepts in the study of the physical world? These are worrying questions. After all, infinite sets arose in human intuition too.

But constructivism does have something to teach us about the mathematical character of Nature. Gödel taught us that there must always be some statements of arithmetic the truth of which we can neither prove nor disprove; but what about all those statements whose truth we *can* decide by the traditional methods of mathematics? How many of them could the constructivists prove? Can we build, at least in principle, a computer that reads input, displays the current state of the machine, determines a new state from its present one, and then use the computer to decide whether a given statement is true or false in a finite time? Is there a specification for such a 'machine' that will enable it to decide whether all the decidable statements of mathematics are either true or false? Contrary to the expectations of many mathematicians, like Hilbert, the answer turned out to be 'no'. Alan Turing in Cambridge, and, independently, Emil Post and Alonzo Church in Princeton, found statements whose truth would require an *infinite* time for any idealized machine to demonstrate. They are, in effect, infinitely deeper than the logic of step-by-step computation.

For our purposes, a constructivist is a formalist with one hand tied behind his back. By limiting the mathematician to only some of the rules that he had been in the habit of using, the scope of his deductions is reduced. The musicologist could take a similarly ascetic view of music, and conceive of different musics whose rules of composition were limited in different ways. Seen in this light, one can sense the frustration that would be felt by the composer who was allocated the most restrictive set of rules and the fewest notes. He could do nothing that other composers could not do, but there would be much that they could do that was beyond his reach. This is the feeling that many mathematicians have about constructivisim. Undoubtedly, some mathematics is constructed in a formal way; but there seems no reason to believe that it all needs to be. An outstanding question is whether all the mathematics needed to describe the physical Universe is within the reach of the constructivist.

The sound of silence: decomposing music

> ... there is no art without constraint. To say that music is an art is to say
> that it obeys rules. Pure chance represents total liberty, and the word *con-*
> *struct* means precisely to revolt against chance. An art is exactly defined by
> the set of rules it follows. The role of aesthetics considered as a science, is to
> enumerate these rules and link them with universal laws of perception.
>
> Abraham Moles

If we think there is anything in the ancient link between mathematics and music, then we should attempt to place music in one or other of the four philosophical pigeonholes we have just introduced—or perhaps, as is often best with mathematics, put some music in one category, some in another. We have concluded that the Platonic (absolutist) view of music seems unnecessarily metaphysical. It asks us to believe that composers discover music, rather than invent it. Now, whereas a Platonic view of mathematics can point to other pieces of evidence in its support—the way in which pure mathematics, derived long ago, so often turns out to give a precise description of some part of the physical Universe, for instance—the Platonic philosophy of music, despite its antiquity, has little to commend it. It suffers from all the weaknesses of the Platonic view of mathematics, but possesses none of its strengths in mitigation. Nature does not display musical structure; musical creations are not culturally independent; nor do they turn out to have vast unforeseen layers of structure that link them to other formally distinct musical creations. Music may be generated; it may be invented; but it is surely not discovered.

Whereas there are common factors linking the mathematics developed by different individuals in different cultures, music implies quite the opposite. Its patterns and rhythms differ significantly from culture to culture; its functions are the common factors. In the Muslim countries of North Africa and the Near East, there is little instrumental influence in music. It is monophonic, dominated by the singing voice, and distinctly unmelodious to Western ears. In southern Africa, the style changes again, to many-layered rhythms contributed by many performers. All this diversity argues persuasively against a Platonic view of music, without ever raising an objection that the better-founded Platonic view of mathematics must face: how do we make contact with this other world of musical forms? Whereas we would expect to communicate in some way with extraterrestrials using the language of mathematics, we would not expect to make much progress using music.

Plato's own view of music, like his opinion of the other fine arts, was to regard them as pale reflections of the ideal unseen forms of universal harmony. His interest in music was largely confined to the ethical harmonies that might flow from its performance—an appreciation of which might bring us closer to the harmonious ideal world from which its structure was drawn. By contrast, Aristotle, Plato's more pragmatic pupil, realized that the pleasure that music brings was a thing of value that owed something to the impression of the performer's personality upon it. The ideal forms were not enough to explain all the individual facets of the music we encountered. And, even if they were, would we really have explained anything? We would be left with a Platonic heaven filled with musical forms, whose harmonious features would still need explaining.

Formalism and constructivism differ as views of mathematics because there are forms of mathematical deduction that cannot be reduced to step-by-step deductions from the natural numbers 1, 2, 3, 4, . . . That is, there are deductive steps that a computer could not carry out in a finite time. This possibility does not exist in musical composition, and so a formalist philosophy of music is, in practice, a constructivist one. That is, it assumes that there exists a set of musical building-blocks—notes, intervals, and so forth—together with the set of rules for combining them to produce phrases, melodies, and so on. 'Music' is the set of all possible applications of the rules to the building-blocks. Considerable progress in exploring an analysis of this sort has been made by the pioneering work of Christopher Longuet-Higgins at the University of Sussex. He has isolated many of the essential structural features that are embodied in classical Western musical composition, laying great stress upon the timing-structure that expert performers introduce into their performance, making it individual to them and attractive to the listener. The success of this isolation of defining features of attractive music can then be tested by programming a computer to compose and perform according to the same principles. The rationale of this production of computer-generated music is not to replace human performance, but to use the nuances of musical composition and performance as a formidable test of attempts to create forms of artificial intelligence. If we could understand what the brain does in music-making, we would have discovered something fundamental about its workings.

Harmony exists because certain combinations of notes are judged more pleasing than others. A theory of harmony has to describe them, and explain why some seem more natural than others. Longuet-Higgins has argued that a simple model can be used for the attribution of musical key. He shows that every interval in music can be represented, in just one way, by a combination of three variables: octaves, perfect fifths, and major thirds. Part of the infinitely repeating tonal space of major thirds and perfect fifths is shown in Figure 5.4. When a listener hears a musical passage, he attributes a key to it by selecting a region of this space. Within a given key, one can ignore the dependence on octaves and treat the tonal space as being two-dimensional, as shown in Figure 5.4. If the choice of key results in the listener having to make large jumps in the table, then he abandons the choice and selects another region of the table (that is, a different

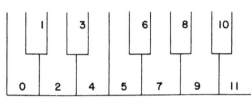

F♯	C♯	G♯	D♯	A♯	E♯
D	A	E	B	F♯	C♯
B♭	F	C	G	D	A
G♭	D♭	A♭	E♭	B♭	F
E♭♭	B♭♭	F♭	C♭	G♭	D♭

Major thirds

y ↑

Perfect fifths

x

6	1	8	3	10	5
2	9	4	11	6	1
10	5	0	7	2	9
6	1	8	3	10	5
2	9	4	11	6	1

5.4 A representation of the tonal space underlying the attribution of musical key proposed by Christopher Longuet-Higgins. Within a given key all harmonic intervals are specified by a two-dimensional array of perfect fifths and major thirds. Within this array each note is pitched one perfect fifth higher than the note on its left, and one major third higher than the note immediately beneath it. Therefore, if we mark the notes within any given key, they appear in groups of adjacent notes, and these groups have different shapes according to whether the key is minor or major. Modulations between keys use the fact that any two keys will have at least one note in common (C major is outlined in the figure). A listener attributes a key to a passage of music by selecting a region of the array. If this choice results in having to make large jumps within the chosen region, it is abandoned and another region is selected in which the tones are more economically clustered, and a new key is attributed.

key) where the sequence of tones can be represented more compactly. The notes of any scale occur in neighbouring clusters, whose shapes are determined by whether the key is major or minor. From this simple pattern, Longuet-Higgins is able to display all the ways in which composers can modulate from one key to another with at least one shared note.*

---

* Interestingly, the structure is that of a mathematical group. It is not large enough to be equivalent to the whole of arithmetic, but seems to match the structure of an arithmetic in which only addition and subtraction are included (no multiplication and division). This smaller 'Presburger' arithmetic is decidable and does not display the domain of arithmetics which must display Gödel incompleteness.

There are many ways in which a given set of musical notes can be performed. Some sound attractive to the ear; others do not. This means that the rules needed to generate interesting music by artificial means—and therefore to define it uniquely as a logical formalism—need to be far more extensive than the usual ingredients of even the most detailed musical score. However, the inability of the ear to discriminate sounds that are too close in pitch and intensity keeps even these unwritten rules to a finite number. In mathematics, the rules governing the permitted logical steps are unambiguous and easily stated. If we could map out in front of us the vast sea of mathematical deductions that follow from all the possible starting assumptions, or axioms, then many of those statements would be devoid of interest to mathematicians. They would be logical deductions, none the less, and hence part of mathematics as defined. However, the musical version of this situation finds the realm of music dominated by a vast cacophony of sounds that are not 'musical' in the conventional sense. The rules for placing the next note are not in practice well defined in the way that mathematical logic is strait-jacketed. One could make them so, but there are many ways in which this could be done; each would produce a different definition of music and a catalogue of sound sequences that the discerning ear could readily distinguish. There is no 'rule' for generating the next note in a piece of music that depends only upon the last note, or even upon all of the notes played so far. Thus, the formalistic picture of music as the set of all possible sequences of sounds that develop from all the possible first notes, using all possible developments, fails to capture what distinguishes music from noise.

If we were able to scan all possible sequences of musical symbols, we would find almost all of them to be random—in the sense that no abbreviation of them could convey all their musical information to someone else. Most sequences of numbers are patternless and random in this same sense. They cannot be abbreviated by replacing their information content by a briefer rule, a formula, or some other mnemonic. None the less, there have been attempts to generate all possible musical sequences—within certain limits. Mozart once wrote a waltz which gave eleven possible variations for fourteen of the sixteen bars of music, with a further two options for the performance of one of the other two bars. This gives 2×11^{14} possible waltzes—enough to keep a million dance-a-day couples occupied for two million years! More recently, David Mutcer, a Harvard professor of electrical engineering, programmed a synthesizer to generate, systematically, all the 50-note melodies that can be created by selecting each note from the 88 on the piano keyboard. The time at which each note was to be played was decided by a random number generator. The computer then began listing all the possible sequences of fifty notes. Eventually, all 88^{50} possible melodies would be listed. This number is stupendous—there are 'only' about 88^{41} atoms in the visible Universe—and in the course of the experiment only a minute fraction of possibilities were generated. As one might expect, the vast majority of the 'melodies' produced so far by Mutcer's Musical Machine are indistinguishable from noise, though a pleasant and vaguely familiar tune will occasionally emerge. But even with a known 50-note melody, the computer-generated version will sound flat and uninteresting to most ears, because the generation process allows no variation in the intervals between notes, and excludes all the harmonic possibilities that are added when multiple tones, rather than single notes, are sounded at any one time.

The imitation game: listlessness

Ultimately, all confusion of values proceeds from the same source:—neglect of the intrinsic significance of the medium.

John Dewey

There is a useful way to grade the attributes of things that we find in the world. The simplest attributes are those for which a definite procedure exists to determine whether or not something possesses it. Human beings can often perform this test unaided by machines; for example, we can tell whether an object floats in water, or whether a given number is even or odd. Tests for some attributes, while straightforward in principle, are extremely laborious to conduct: for example, we can in principle always tell whether a number is prime or not; but if the number is large, we shall need help from a fast computer; and if the number is very large (with thousands of digits), then even our fastest computers might take thousands of years to tell. None the less, in principle, the check could be carried out for any given number, and the answer found to be 'prime' or 'not prime'. On reflection, we see that much of our education system is dedicated to teaching young (and not so young) people to detect the presence or absence of attributes in this manner: 'Is this a verb?'; 'Is this sentence grammatically correct?'; 'Is this triangle equilateral?', and so forth. We have become so accustomed to the technological solution of our problems that it is easy to get lulled into thinking that we can decide the presence or absence of any attribute in a similar fashion, just by building faster computers. This is far from the case. Indeed, it is not possible even to decide the truth or falsity of all statements of arithmetic by implementation of a computer program.* Thus, there are attributes of the world whose truth or falsity cannot be decided by the application of a test that takes a finite number of steps to implement.

Another property one can ask of an attribute of the world is that it shall be 'listable'—that is, is there a definite procedure that lists all the examples

---

* It is, however, possible in the case of all statements of Euclidean geometry.

possessing the attribute? This list might be infinite (as would be the case if the attribute was something like being an even number), in which case the listing process would continue indefinitely. 'Listability' differs from decidability because, although an attribute may be listable, there may be no way of listing all the entities that do not possess the attribute in question. The problem of deciding whether this page is written in correctly spelt language is a decidable one. The page contains a finite number of words, and they can each be compared against dictionary definitions of spelling in all tenses and cases. (This is what a word processor's 'spell-checker' does.) Every word can be judged correct or incorrect by that (or any other) criterion. Nevertheless, this page of immaculately spelt words could still be gibberish in any known language. If, however, a grammar-checker were to pass the page of words as grammatically correct, the writing would still remain meaningless to a reader who knew nothing of the language in which it was written. As the reader learnt some of that language, so parts of the page would become meaningful; but we could not predict which parts would become intelligible; nor could we predict whether the reader would ever write an identical page of words in the future. The property of being an intelligible page of language is thus listable but not decidable.

Unfortunately, truth is neither a listable nor a decidable property; nor is the truth of a statement of arithmetic. The American logician John Myhill has used the term 'prospective' to characterize those attributes of the world that are neither listable nor decidable. They are properties that cannot be recognized by the application of some formula, made to conform to a rule, or generated by some computer program. They are characterized by incessant novelty that cannot be encompassed by any finite set of rules. 'Beauty', 'ugliness', 'truth', 'harmony', 'simplicity', and 'poetry' are names we give to some of the attributes of this sort. There is no way of listing all examples of beauty or ugliness, nor any procedure for saying whether or not something possesses either of those attributes, without redefining them in some more restrictive fashion that kills their prospective character.

This division of the attributes of the world into those with decidable, listable, and prospective properties helps to clarify where attempts to impose philosophies of mathematics upon music fall down. We could list all possible sound sequences generated by a prescribed list of instruments playing alone or in unison; but we could not implement a universal criterion for deciding whether they would sound harmonious or not; nor could we write a program that would generate the subset of all those sound patterns that were 'harmonious'—let alone 'meaningful'—to the human listener. Musical appeal is a prospective property. It looks as if it might be listable or decidable only because, like words on a page, music is written down using a finite number of symbolic marks on pieces of paper. But that prescription is necessarily incomplete, and much of the attractiveness of music is added in the special translation process that we call performance.

The inventionist philosophy is an implausible explanation for the whole of mathematics because it fails to account for the unreasonable effectiveness of mathematical descriptions of Nature—descriptions that are more impressive the further one moves from the phenomena of immediate and past human experience. An inventionist philosophy of music is more persuasive. It views music simply as an activity of musicians. Its character is universal only in respect of certain psychoacoustic elements associated with physiological or neurological features common to human listeners, or by appeal to the universal properties of sound. In other respects it reflects the diversity of human cultures, of social trends, and of our reactions to those trends.

▓▓ The sound of music: hearing and listening

> We are reluctant, with regard to music and art, to examine our sources of pleasure or strength. In part we fear success itself—we fear that understanding might spoil enjoyment. Rightly so: Art often loses power when its psychological roots are exposed.
>
> Marvin Minsky

The adaptationist explanation for the advent and successful maturing of an ability like musicality ascribes its ubiquity to the fact that it is advantageous, on balance, for humans to possess it. Alternatively, one might place greater emphasis upon the instinctive aspects of a mental ability, and seek to show that it is primarily fashioned by natural selection, rather than acquired by learning or as a byproduct of genetic programming for something else.* In contrast, most social psychologists seek to attribute human abilities to the particular social context within which individuals develop, or to repeated interaction between individuals. The social scientist might see musical style and content as an outcome of specific human concerns or economic constraints. From another viewpoint, a physicist might treat musical harmony simply as a sonic phenomenon received by a frequency analyser (the ear) connected to a computer (the brain) that is sensitive to structured pulses of sound within prescribed ranges of frequency and intensity. A further branch of study, 'psychoacoustics', would be needed to discover the relationship between the principal physical properties of the sound—its frequency, intensity, or spectral variation—and the qualities of pitch, loudness, and timbre perceived by listeners. In the rest of this chapter we

---

* Of course, even 'innate' abilities must have an origin and associated *raison d'être*. Their initial structure must arise either by pure chance, through selection acting upon alternatives, or because they constitute the unique design that will achieve a particular beneficial effect.

shall see what light can be shed upon the nature and influence of musical sound by taking the viewpoint of the physicist. This will help us to isolate which properties of music are rendered inevitable for us by physiological and neurological features of the human condition.

A work of art should display order at some discernible level—preferably at many levels. This ordering means that there is a pattern and a set of rules for combining sounds, or colours, according to the medium employed to represent the pattern. In the case of music, the results can be viewed in four ways: in terms of the raw materials used, the sounds that convey the music, the psychological responses to them, or the information content of the music. An understanding of what music 'is' requires a discussion of all these aspects. No single one of them can give the whole picture, yet each offers important insights. For instance, we can study music as an acoustic phenomenon to discover if emotionally appealing music possesses common features; then, by relating those properties to our perceptive apparatus, we might discover why some acoustic patterns produce strong psychological responses.

We should begin by putting music in a broader acoustic context. What we call the 'pitch' of a sound is determined by the frequency of the vibration it excites within our ears. When sounds have frequencies lower than about 16 cycles per second* we cease to hear them and begin to feel them as vibrations in our surroundings. This very-low-frequency domain is called the infrasonic region.† Above 20 kHz, sounds enter the ultrasonic region—again, beyond the range of our hearing, although young children can generally hear slightly higher frequencies than adults. Many animals, like cats and dogs, can hear far higher frequencies: up to 60 kHz in the case of cats. Yet the 1250-fold range of sound frequencies to which the human ear is sensitive dwarfs the tiny, twofold range of light frequencies that the human eye can detect. The far greater density and quality of the information that the visual sense processes is enormously expensive in terms of the brain's resources. Extending those visual abilities over a much wider frequency range would not have represented the optimal utilization of mental resources in an environment that was in darkness for half the day.

After frequency, the most important property of sound is its intensity level: how loud it sounds. Again, human physiology determines which sound levels we can hear. The lower threshold of human audibility defines the zero decibel level, and sounds above about 130 decibels are intense enough to produce deafness. These numbers require some further explanation. Sound intensity levels are

---

* One cycle per second is also called a 'hertz' and denoted by the abbreviation Hz; one thousand Hz is denoted as 1 kHz (one kilohertz).

† An interesting comparison can be made between this low-level threshold and the frequency of so-called 'alpha' brain waves, at 10 Hz, which result when you close your eyes and think about something non-visual. They stop if you open your eyes.

commonly measured in 'decibels', where a decibel (abbreviated dB) is defined to be ten times the logarithm (to the base ten) of the sound level in units of an intensity level of 10^{-12} Watts per square metre. This sounds rather Byzantine, but it is defined like this so that one decibel is equal to about the faintest sound that a normal person can hear. Hence, a sound intensity that is a thousand times the base level would correspond to 30 dB. It helps to compare these numbers with more familiar sound levels: the rustling of the trees in the breeze, as you stroll through the woods on a spring day, produces about 10 to 18 dB; an orchestra produces between 40 and 100 dB; ordinary conversation produces about 65 dB, but a whisper little more than 16 dB; rush-hour traffic can generate 70 dB; enthusiastic hammering, or a clap of thunder, creates about 110 dB. Our lower threshold of audibility witnesses to an extraordinary sensitivity. The quietest audible sound at a frequency of 1000 Hz is the result of the ear's inner membrane being displaced by one-tenth of the diameter of a hydrogen atom. This is only a little above the sound level created by the continuous buffeting of the eardrum by air molecules at everyday temperatures.* In Figures 5.5 and 5.7, the domains of loudness and frequency that are accessible to the ear are mapped out, together with the regions employed in music.

The perceived sound of music depends delicately upon the architecture of the ear. Like our other sense organs, the ear is a structure of extraordinary complexity. The eardrum is a thin membrane dividing the middle and outer ears. It remains in contact with air at atmospheric pressure on either side via the Eustachian tube (see Figure 5.6). An incoming sound wave creates a succession of compressions and rarefactions in the air within the auditory canal of the outer ear; this produces pressure variations across the eardrum, causing it to vibrate back and forth. These vibrations are transmitted by a chain of tiny bones, along the middle ear, through an opening into the inner ear, where they disturb a fluid that passes the disturbances on through the cochlea. Next, they disturb the basilar membrane, whose movements are registered by tiny hair cells, which are able to transmit those signals to the central nervous system where, ultimately, they register the sense that we call 'hearing'. These signals are sent only when the incoming vibrations have frequencies in the 'audible' range, 16 Hz to 20 kHz. We perceive the frequency of these oscillations as their 'pitch'; their amplitude, which increases with the magnitude of the pressure variations in the air within the auditory canal, is sensed as 'loudness'. The ear does not respond equally to all incoming frequencies within the audible range: it will perceive sounds with the same intensities, but different frequencies, to have slightly different loudnesses.

* If you place a sea-shell to your ear, the rushing 'sound of the sea' that you then hear is the sound of your bloodstream. The shell screens out the background noise that normally predominates and renders it inaudible. A similar effect can be heard when the hearer is inside a sound-proof room or subterranean cave.

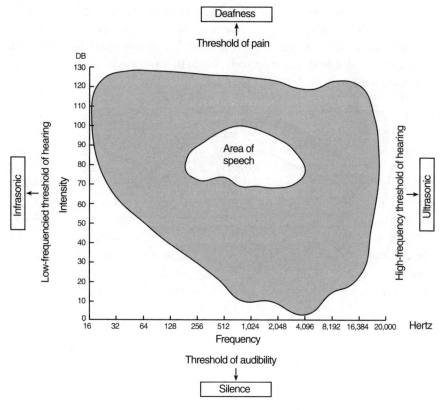

5.5 The audible region within the domain of sound intensity (in decibels) and frequency (measured in hertz).

An interesting feature of music, like many other examples of complex emergent phenomena, is the way in which it has evolved to fill the intensity–frequency domain available to it. History shows that music has been getting steadily louder and more diverse in its range of pitch. Before the Renaissance, musical pitch of the fundamental frequencies ran from about 100 to 1000 Hz, and mirrored the frequency range of the human voice. As new instruments have been added to the orchestral repertoire, this range has steadily expanded. The advent of synthesized electronic music means that there are now virtually no barriers to the frequencies (or intensities) at which musical sounds can be generated. A comparison of the domains of pre-Renaissance music and nineteenth-century orchestral music is shown in Figure 5.7.

Each instrument has a comparatively narrow dynamic range—far smaller than that of the orchestra as a whole, or that of the piano, which spans the greatest frequency range, as shown in Figure 5.8.

Our hearing has evolved so as to interpret pitch changes rather than absolute

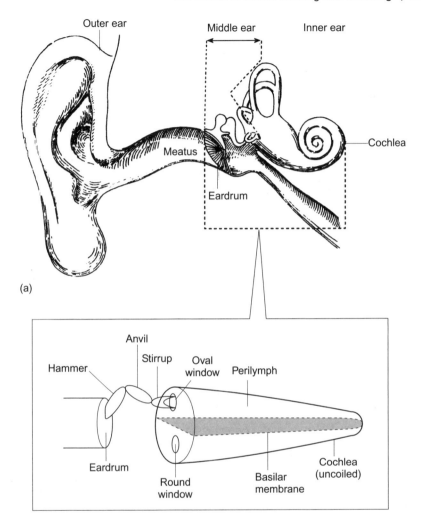

5.6 (a) The human ear; (b) fine detail of the middle and inner ear showing the components that transmit vibrations of the eardrum down the chain of bones (hammer, anvil, and stirrup), through the oval window, where disturbances are created in the perilymph fluid, which sets the basilar membrane in motion. This motion is picked up by hair cells, whose response sends signals to the nervous system. Low-frequency sounds activate hair cells at the far end of the membrane; high frequencies excite only cells near the round window area.

pitch levels. It has proved more economical to invest neurological resources in sensing *changes* of pitch, rather than in developing the more sophisticated calibration that is needed for absolute pattern-recognition. Some people nevertheless

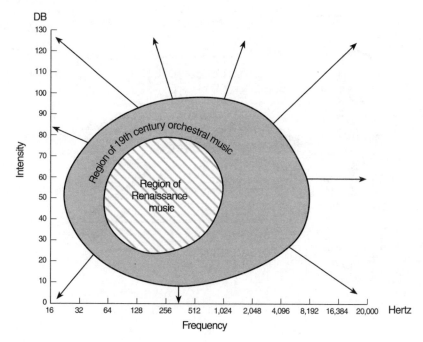

5.7 The ranges of sound intensities and frequencies employed in Western music have evolved to fill a greater fraction of the whole audible range. Here the domain of Renaissance music is compared with that of nineteenth-century orchestral music. Modern electronic music can, in principle, be designed to fill the entire audible domain of Figure 5.5.

have the ability to recognize, or produce, notes at an absolute pitch. This much-admired ability is called 'perfect pitch'. Our ears are sensitive to changes in frequency of merely one-half of one per cent in the audible range. The brain makes little or no long-term use of information about absolute pitch levels. Most of us remember this information for only a few minutes. It is not known whether one could teach very young children to have perfect pitch recognition in the early stages of their mental development, so that pitch information eventually became stored in long-term memory as well.

A further curiosity of our sensitivity to changes of pitch is how it is under-used in musical sound. Western music, in particular, is based upon scales that use pitch changes that are at least twenty times bigger than the smallest changes that we could perceive. If we used our discriminatory power to the full, we could generate an undulating sea of sound that displayed continuously changing frequency rather like the undersea sonic songs of dolphins and whales.

When the brain receives sequences of musical tones, it does what it does with other patterns: it attempts to 'interpret' them by sensing the smallest number of

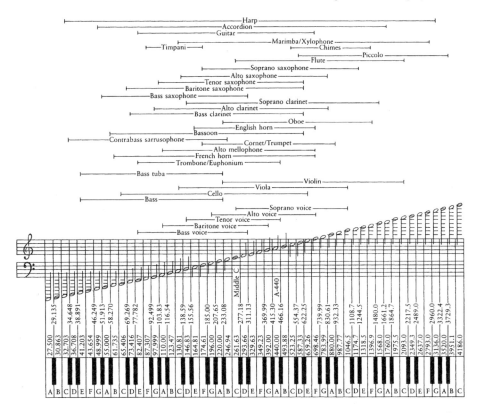

5.8 The frequency ranges of modern musical instruments and human voices compared with the range of the piano keyboard.

cues within the signal. If this strategy fails to identify the signal, the brain then makes use of information stored in its long-term memory about previous, similar experiences. This information may allow some aspects of a future signal to be anticipated—as it does when we hear the first line of a familiar song. This ability to extrapolate forwards on the basis of past experience is one form of that ability that we call 'intelligence'; it can dramatically enhance an organism's chances of survival. The alternative way of dealing with its environment is by instinctive reaction. Instincts are preprogrammed responses to well-defined situations; unlike learned responses, they cannot be continuously updated. Instinctive responses are much simpler and more economical in their use of neurological programming and resources. They are 'cheaper' to develop than learned behaviours, but are much more likely to prove disadvantageous when novelties are encountered—as, say, in the form of rapid environmental change, or new forms of competition. Nevertheless, instinctive behaviours should not necessarily be regarded as totally second-rate. Our most remarkable attribute, that of language, appears to be of this instinctive type.

There have been studies to identify the brain's response to the absence of an expected stimulus. A lacuna still results in a response similar to that commensurate with the expected stimulus. Some re-evaluation procedure then comes into play when the expected stimulus is found to be absent. We might associate this with the tension that is created when a musical development follows new or unexpected avenues, or when a sound is unexpectedly discordant. A very complex musical work will stimulate the auditory nerves, and hence the brain, to produce matchings and extrapolations in enormous numbers, at a very great rate. The fact that music-lovers have a pleasant experience upon hearing the same piece of music on many occasions suggests that this neurological response occurs automatically whenever the music is heard. Those who get little, or no, pleasure from music may have auditory nervous systems that can handle information of this type only at a slower rate, and so the entire experience is barely, if at all, stimulating. When some subtle change in the sound occurs that stimulates the music-lover, and perhaps produces some other emotion as well, the less-musical listener's sound processors are already saturated by the underlying flow of musical information; hence this new subtlety elicits no response— even though, at the purely acoustic level, it is heard.

The perception of musical sound is also influenced by the arena in which it is heard. This environmental effect is familiar to us: we all think we sing rather well in the bath, but not so well in the open air.*

Although we have been building auditoria since the days of the early Greeks, 2500 years ago, their acoustics were not fully understood, in a way that enabled them to be optimized for musical performance, until the early years of the twentieth century. Even today, plans to improve the acoustic qualities of a concert hall may well find themselves compromised by the needs of structural safety, size, cost, and architectural appearance. Although many factors combine to determine the quality of sound an audience hears in a building, the most important characteristic of an auditorium is its *reverberation time*. This is a measure of how rapidly any reflected sound decays away into inaudibility. More precisely, it is the time required for sound to decrease a million-fold in intensity

---

* Sound reflects well from the hard, smooth walls of the shower room, and there is considerable reverberation, which makes your sound-level rival Pavarotti's. Moreover, many natural frequencies of vibration are available within the air along all three perpendicular directions between the two pairs of facing walls, and between the floor and ceiling (whereas a stringed instrument can take advantage only of waves along the direction of the string). These can be excited by a singer. Many of these frequencies lie close together, within the frequency range of the human singing voice; the bathroom singer thus receives impressive background support from many naturally occurring resonances.

(that is, by 60 dB). A good concert hall will have a reverberation time of about two seconds.

It is important that sound decays smoothly, at a constant rate. If the decay were sporadic, or occurred rapidly for one second, and slowly thereafter, then the music would sound very uneven. Our auditory system would simultaneously receive sounds that were generated at different times, and would have the formidable problem of reconstituting their original ordering in the face of varying decay times: we would not then be able to unravel them cleanly enough to reconstruct a smooth musical statement.

To understand why there are optimal reverberation times for different varieties of sound, we need to be aware of the fact that roughly three-quarters of the intensity of the sound will be dissipated in one-tenth of the reverberation time; after this amount of attenuation, the ear is ready to distinguish a new sound. One-tenth of the reverberation time therefore gives, inversely, the number of new sounds that the ear can comfortably resolve per second. This reveals that we should design a lecture theatre to have a reverberation time of about half a second, so that an audience will perceive distinct new sounds at a rate of about twenty per second—a good match to the rate of production of new sounds by a human speaker, and of their reception by a human listener. But such a theatre would be a poor auditorium for music. Most music sounds best in halls with reverberation times of about two seconds, thus providing listeners with about five new sounds per second—close to the rate of performance of notes in many forms of music. If the reverberation time is too long, then the sound becomes confused because the audience hears simultaneously too many notes produced at different times. But if the reverberation time is too short for the sound being heard, then each sound is heard standing alone, rather than as part of a continuous musical statement. Some reverberation times for different types of enclosed building are shown in Figure 5.9.

Acoustic engineers attempt to predict the acoustic performance of concert halls by using computer simulations to discover how the acoustic properties depend upon features like the reflectivity of the walls, the size and shape of the hall, or where you happen to be seated within it. These quantitative studies were pioneered by a remarkable American scientist, Wallace Sabine, whose interest in such problems was aroused when, in 1895, Harvard University asked him to discover why lecturers at the university's newly opened lecture room in the Fogg Art Museum were proving unintelligible to their audiences. Displaying an admirable confidence in the lucidity of the Harvard professoriate, Sabine began painstaking detective work using different sound sources and a stop-watch. He found that the very long reverberation time of the room (5.62 seconds almost everywhere) was responsible for making the students think that they had enrolled at the University of Babel. By placing sound-absorbent padding on

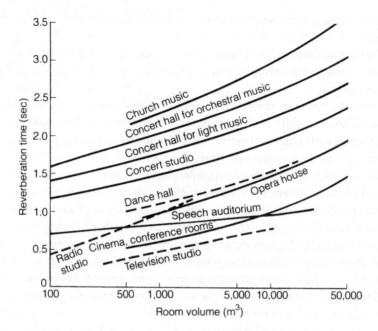

5.9 The best reverberation times for buildings of different volume. The optimum reverberation time depends upon the type of sound being produced. Excessive reverberation occurs when the auditorium is too large—usually because the ceiling is too high—or where the interior surfaces reflect incident sound too readily.

some of the walls, and cushions on the seats, he reduced the persistence of the lecturer's voice by lowering the reverberation time to about 1.1 seconds, and the university's problem was solved. Sabine went on to greater things. In 1900 he provided the acoustical design for the new Boston Symphony Hall—still said by many to be, acoustically, the finest concert hall in the world. He determined the optimal reverberation time for music by conducting trials with a variety of musicians and trained musical listeners to arrive at some unanimity, and designed accordingly.*

Different styles of speaking and singing are best heard in auditoria with optimal reverberation times, which vary with room volume for different types of structured sound-generation; some are shown in Figure 5.9. These trends shed light on the link between the evolution of musical styles over the centuries and

---

* There are many other architectural factors that influence our perception of music. If, for example, the delay until players hear the sound they are producing, and that of their colleagues, reflected back at them from the walls is too long, then they begin to feel they are not playing together in an intimate environment, and their performance suffers. Delay times greater than two-hundredths of a second are noticed as disturbing; the finest concert halls give between two-thousandths and nine-thousandths of a second delay.

the nature of the buildings within which the music was performed. Much choral and organ music exemplifies the slow, majestic sound that is best heard in buildings like vast cathedrals, with long reverberation times. Architectural factors therefore enhance the transcendental atmosphere of this musical element of religious worship.

By contrast, music of the Baroque period (1600–1750), reaching its climax with Bach, was generally composed for performance in smaller halls, theatres, and churches, with highly reflective walls. These environments possess greater intimacy, and have comparatively short reverberation times: at or below about 1.5 seconds. Much of the stylistic variation displayed by music of this period is a reflection of the wide range of sites, with different reverberation times, in which its performance was intended to sound crisp and clear. During the ensuing Classical period (1775–1825), the make-up of the concert orchestra evolved into its present form, although music was generally performed in concert halls rather smaller than those used today—typically with reverberation times near to 1.5 seconds, rising to about 1.8 seconds in the largest nineteenth-century auditoria. This is one reason why classical music now sounds best in concert halls with this narrow range of reverberation times. By contrast, late Romantic composers, like Tchaikovsky and Berlioz, require longer reverberation times for the emotional effect of their music to be fully felt. It is not surprising to learn that these works were written at a time when they could be performed in the first large concert halls with two-second reverberation times.

Adventures of Roderick Random: white noise, pink noise, and black noise

Of all the noises, I think music the least disagreeable.

Samuel Johnson

The world around us is full of patterns: of light, of sound, and of behaviour. As a result, the world finds itself well described by mathematics, because mathematics is the study of all possible patterns. Some of those patterns have concrete expressions in the world around us—where we see spirals, circles, and squares. Others are abstract extensions of these worldly examples; yet others seem to reside purely in the fertile minds of their conceivers. Viewed like this, we see why there has to be something akin to 'mathematics' in the Universe in which we live. We, and any other sentient beings, are at root examples of organized complexity: we are complex stable patterns in the fabric of the Universe. If there is to be life in any shape or form, there must be a departure from randomness and total irrationality. Where there is life there is pattern, and where there is pattern there is mathematics. Once that germ of rationality and order exists to turn a chaos into a cosmos, then so does mathematics. There could not be a non-mathematical Universe containing living observers.

None the less, there could perhaps have existed only a smidgen of order at the heart of things. The part of reality that intersects our own parochial evolutionary history might be a mere drop in an ocean of universal irrationality. Alternatively, the order behind the world, and hence the mathematics needed to describe it, might be of a deep and uncomputable kind. The patterns of Nature might be unfathomable by any living subset of those patterns. If the mathematical structure of such a world were permeated by the elusive, uncomputable functions of Turing, then mathematics would not help its inhabitants to predict the future, explain the past, or capture the present. But, again, perhaps such worlds could not be inhabited by sentient beings. In order for such creatures to survive in a complex natural environment, some regularities must exist within that environment, and preconscious minds must be able to embody some of those environmental regularities. For living complexity to evolve successfully it must be able to store representations of its environment, and carry out computations of steadily growing complexity. The success of this process relies upon a bedrock of reliable pattern that can be approximated step by step. A world governed by uncomputable mathematical structures does not permit life to evolve by a succession of small variations, each producing an improved adaptation to reality. Such worlds would lack life. Seen in this light, the existence of a certain level of discernible order in the natural world is neither unexpected nor mysterious: at least, no more—and no less—so than is our own existence.

Faced with a conclusion of this sort, we need to look more closely at the natural patterns around us. Art-forms like music are patterns too, but they seem to have little in common with Nature. Music does not sound like an imitation of anything. But if we have evolved to cope with the changing patterns of a complex environment, there may be naturally occurring forms of complexity that our brains are best able to apprehend. In such circumstances, we might expect artistic appreciation to emerge as a byproduct of those evolutionary adaptations that are accommodated to vital natural patterns of variation. The music that we find attractive might therefore share some features displayed by natural patterns of sound.

In order to find the best way of classifying natural and artificial patterns of sound, it helps to see how engineers study sound. A useful quantity is the *power spectrum* of the signal, which displays how the average behaviour of a time-varying quantity varies with frequency. If the sound is oscillatory, then the average is taken over a time much longer than the period of a single oscillation (typically over more than about thirty oscillations). Another informative quantity, the *auto-correlation function* of the signal, is a measure of how the signals at two times, t and $(t + T)$, are related. If the signal is on average the same at all times, then the autocorrelation function will not depend on the absolute time t, but merely on the interval of time, T, between different observations of the signal.

Sound-sequences defined by a power spectrum are what physicists and engineers call 'noise'. An important feature of the power spectra of many natural sources of noise is that they are proportional to a mathematical power of the signal frequency over a very wide range of frequencies. In this case, there is no special frequency that characterizes the process—as would result from repeatedly playing the note with the frequency of middle C, for instance. Such processes are called *scale-free*. If one halves or doubles all the frequencies, then a scale-free spectrum would keep the same shape. In a scale-free process, whatever happens in one frequency range happens in all frequency ranges. If music were exactly scale-free over its entire frequency range, then a gramophone record would sound the same at any playing speed (if compensatory changes in volume were made). Obviously, the human voice is far from being scale-free over the whole frequency range of normal conversation because we know that a speeded-up broadcast of the human voice sounds distinctly like Donald Duck.* Likewise, a cello or a violin sounds quite different when speeded up or slowed down; by contrast, pure scale-free noises would sound the same.

Scale-free processes have power spectra that are proportional to inverse powers of the frequency f, as f^{-a}. The character of the noise changes significantly if the value of the constant a is altered. If noise is entirely random, so every sound is completely independent of its predecessors, then a is zero, and the process is called *white noise* (see Figure 5.10a). Like the spectral mixture that we call white light, white noise is acoustically 'colourless'—equally anonymous, featureless, and unpredictable at all frequencies, and hence at whatever speed it is played. It has zero autocorrelation. When your TV picture goes haywire, the 'snow' that blitzes the screen is a visual display of white noise that arises from the random motion of the electrons in the circuitry. At low intensities, white noise has a soothing effect because of its lack of discernible correlations. Consequently, white-noise machines are marketed to produce restful background 'noise' that resembles the sound of gently breaking ocean waves.

The lack of any correlation between samples of white noise at different times means that its sound-sequence is invariably 'surprising', in the sense that the next sound cannot be anticipated from its predecessor. By contrast, a scale-free

---

* For spoken English, the speech sounds are statistically uncorrelated, on the average, over frequencies below about 2 Hz, but become correlated like 'brown noise' (see p. 233) at higher frequencies. Of course, a close correlation between ideas expressed in spoken words does not necessarily arise from the use of sounds whose frequencies are correlated. A long succession of very weakly correlated sounds may produce a message with significant long-time semantic correlations. There are, however, many non-Western languages in which the pitch-variations of the voice do play a significant role in endowing meaning.

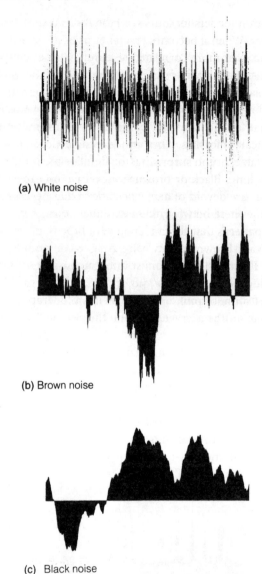

(a) White noise

(b) Brown noise

(c) Black noise

5.10 Samples of (a) white noise, (b) brown noise, and (c) black noise.

noise with $a = 2$ produces a far more correlated sequence of sounds, called *brown noise** (see Figure 5.10b). Brown noise is rather unenticing to the ear; its high

---

* This colourful terminology arises because the archetype for a statistical process of this sort is the diffusion of small particles suspended in a liquid, first observed by the Scottish botanist Robert Brown in 1827, and thereafter dubbed 'Brownian motion' in recognition of his discovery.

degree of correlation renders its course rather predictable. It 'remembers' something of its history. When a becomes greater than 3, we enter the realm of *black noises*—which are even more correlated (Figure 5.10c). Such processes seem to describe the statistics of a wide variety of man-made and natural disasters—from earthquakes and floods to stock-market plunges and train crashes. The highly correlated appearance of such catastrophes could be taken as a basis for the old adage that 'accidents always come in threes'. None of these 'coloured' noises is aesthetically pleasing. They produce sequences of sounds which are either too predictable, or too surprising, to stimulate the mind's pattern analysis routines for very long. Black or brown noises leave no expectation unfulfilled, while white noises are devoid of any expectations that need to be fulfilled. This suggests that somewhere between these extremes of surprise and dull predictability might lie patterns that contain enough of both to arouse our sensibilities.

Between white and brown noise, when a lies between 0 and 2, lies the realm of 'pink noise' (see Figure 5.11). The most interesting example is the intermediate case where $a = 1$, which is called '$1/f$ noise'* or 'flicker noise' by engineers. The most interesting feature of pink noise is that it is moderately correlated over *all* time-scales and so, on the average, it should display 'interesting' structure over all time intervals.

In 1975 Richard Voss and John Clarke, two physicists in the University of California at Berkeley, analysed a variety of musical recordings and broadcasts by radio stations to see if they displayed any spectral affinity to scale-free noises. Their results were striking. They discovered that a wide range of classical compositions were closely approximated by $1/f$ pink noise over a wide range of frequencies. Likewise, synthetic musical compositions in which both the pitch frequency and the duration of the notes were selected from $1/f$ statistics were

5.11 Sample of $1/f$, or 'pink', noise.

---

* Pronounced 'one-over-eff' noise.

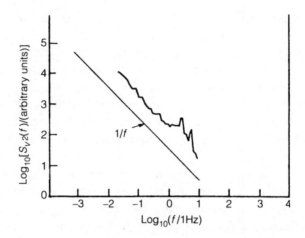

5.12 The spectral density of audio power ('loudness') versus sound frequency, f, in logarithmic units for Bach's First Brandenburg Concerto, measured by Richard Voss and John Clarke.

found to be appealing. In contrast, white and brown noise sources were found uninteresting.

Focusing upon particular musical compositions, Voss and Clarke first studied the audio signal from a performance of Bach's First Brandenburg Concerto. The spectrum, averaged over the whole performance, is shown in Figure 5.12. As can be seen, the spectrum has a slope close to that of $1/f$ noise over almost its entire frequency range. The two sharp peaks between 1 and 10 Hz are associated respectively with the time needed to sound a single note, and the particular musical rhythm used by the composer. Next, Voss and Clarke repeated the experiment for a wider variety of musical sources: some Scott Joplin piano rags, a rock music station, a classical music station, and a news-and-music radio show. Their spectra are shown in Figure 5.13—again averaged over the whole record-ing (or over 12 hours, for the radio stations). The results are striking. There is a strong tendency for all these sources of 'noise' to follow the $1/f$ spectral slope. The Joplin has much more high-frequency (that is, short time-interval) struc-ture around 1–10 Hz than the Bach—a reflection of its distinctive structure—but is still close to a $1/f$ spectrum below about 1 Hz. The average output from the classical, jazz, and rock stations also conforms to the $1/f$ form down to those frequencies that begin to register the typical length of a piece of broadcast music. The news-and-music programme also displays a $1/f$ spectrum, except for inter-ruptions, which pick out the typical time taken for the broadcaster to utter a word (about 0.1 seconds), and the length of a typical item (about 100 seconds). One can also see the effect of the change from white to brown noise, characteristic of spoken English, around 2 Hz.

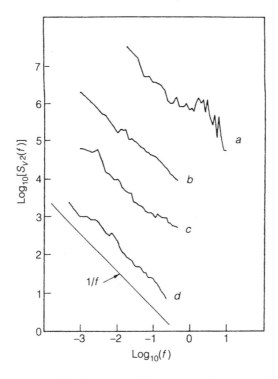

5.13 Loudness variations with frequency, f, for a range of structured sounds: (a) Scott Joplin piano rags; (b) classical music radio station; (c) rock music radio station; (d) news and music radio station; as measured by Voss and Clarke.

One lesson that emerges from these studies is the unrealistic nature of much so-called 'stochastic music', which produces music by programming a random-number generator to select each note—just like Mutcer's melody-generator, which we discussed earlier. This will produce a spectrum resembling that of white noise. Even if some memory of a few previous notes is programmed in to inject some attractive correlations, the result is quite different from that displayed by a $1/f$ spectrum for loudness and the intervals between notes, over a wide frequency range, as shown in Figure 5.14. The $1/f$ 'music' has correlations over all time-intervals; these cannot be reproduced by introducing a single characteristic correlation time, below which the notes are correlated and above which they are not. The single correlation time produces a white-noise spectrum up to some frequency corresponding to the correlation time, but correlations over shorter times then create brown noise at higher frequencies.

Voss and Clarke's work seemed to be an important step towards characterizing human music as an almost fractal process of intermediate complexity in the low frequency range, below 10 hertz. It provoked other physicists with interest

5.14 Examples of musical compositions derived from notes selected for frequency and duration from (a) white noise, (b) 1/ƒ noise, and (c) brown noise spectra.

in sound and complexity to re-examine what they had done in greater detail. It turned out things were not as clear-cut as they had claimed. The lengths of the snatches of music that are used to determine the spectrum of correlations is crucial and an inappropriate choice can bias the overall findings. Work by Nigel Nettheim and by Yu Klimontovich and Jean-Pierre Boon showed that the $1/f$ spectrum was something that would arise for any audio signal recorded over a long enough interval, like that required to perform an entire symphony, or for the hours of radio-station music transmissions that Voss and Clarke recorded. So if you analyse the sound signal for long enough, all musics will display $1/f$ spectral behaviour. This means that Voss and Clarke's analysis of long pieces of music tells us nothing about human musical taste, as they believed. If we go to the other extreme, and consider musical sounds over very short intervals of time that encompass up to about a dozen notes, then we find that there are strong correlations between successive notes and the sounds are very predictable and far from random. This suggests that it is on the intermediate time intervals that the spectrum of music will be most interesting.

Boon and Oliver Decroly then carried out an investigation like that of Voss and Clarke, but confined to the 'interesting' intermediate range of time intervals in the frequency range from 0.03 to 3 hertz. They studied 23 different pieces by 18 different composers, from Bach to Carter, averaging only over each part of each piece. They found no evidence for a $1/f$ spectrum at all. Instead, the spectrum fell as $1/f^a$, with a lying between 1.79 and 1.97. Nettheim had found something very similar based on a study of only five melodies (see Figure 5.15).

These analyses suggest that humanly appreciated music is much closer to the correlated brown noise ($a = 2$) spectrum than the 'pink' $1/f$ noise. There are preferred time intervals in the musical compositions and particular correlations. All these investigations were confined to Western music. It would be interesting to see the results of a study of non-Western musical traditions over the same time intervals.

A philosopher like Immanuel Kant would have explained our affinity for music by appeal to a pre-established harmony between music and the constitution of the human mind. If he were teleported into the present, Kant would not be surprised to find that there are links between the properties of musical sound and the brain's sense receptors. But whereas Kant would regard these links as inexplicable, we have learned to look for ways in which the nature of the environment can gradually imprint affinities for certain patterns of sound because it is advantageous, and hence adaptive, to do so. We suspect that the mind is particularly sensitive to stimuli that exhibit distinctive spectral forms of scale-free noise. A wide range of musical compositions, from a diversity of cultures and musical traditions, exhibit this property. But one should not regard this observation and associated speculation as totally reductionist, any more than

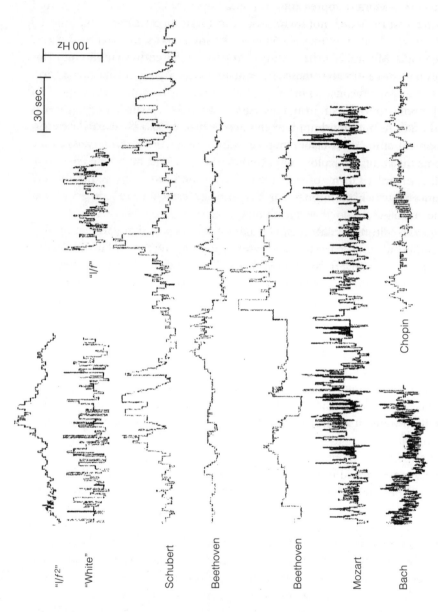

5.15 Spectral representations of various short melodies by classical composers compared with white noise and $1/f$ noise spectra. The lengths of the extracts are 3 or 4 bars, the length of a typical musical phrase, and are taken from work by Nigel Nettheim.

one should take seriously the claims of music-lovers that music is a transcendental art-form whose charms are beyond words. Our minds, their propensity to analyse, distinguish, and respond to sounds of certain sorts, yet ignore others, have histories. Musical appreciation is not an attribute that increases our adaptation to the world: it does not enhance our chances of survival. Were it so, then we would find musical abilities to be widespread amongst other members of the animal world. Musicality seems most reasonably explained as an elaboration of abilities and susceptibilities that were evolved originally for other, more mundane, but essential acoustic purposes. Our aptitude for sound processing converged upon an optimal instinctive sensitivity for certain sound patterns, because their recognition improved the overall likelihood of survival. With the development of the more elaborate processing ability that we call consciousness has come the ability to explore and exploit our innate sensibility to sound. This has led to organized forms of sound that explore the whole range of pitches and intensities to which the human ear is sensitive. Those forms diverge in their stylistic nuances, as do the decorations around people's necks and in their homes, from culture to culture. But the universality of musical appreciation, and the common spectral character of so much of the sound that we enjoy, prompts us to look for the universal aspects of early experience for an explanation. If the nature of our world had allowed us to survive with a very narrow range of sensitivity to different sound frequencies, then our chances of generating interesting music would have been considerably restricted. If our ears had been sensitive only to some interval of ultrasonic frequencies that lie beyond our present ability to hear, then our music would have been concentrated in this frequency range, and our instruments—the devices we use to populate that realm of musical sound—would be very different. Had the sounds that fill our world been different in their spectral properties, then different powers of discrimination would have been needed for us to respond to sounds of impending danger, or to use sound to estimate distances and sizes, and we would need different spectral analysis sensitivities. The result would have been a penchant for sounds with quite different structures—structures that from one perspective would have been more surprising or more predictable.

6 | All's well that ends well

Now, there can be no doubt that perceptual error, if it generally went uncorrected, would prove a biological catastrophe. The man who regularly mistakes his wife for a hat (or worse still, his hat for a wife) is headed for extinction.

NICHOLAS HUMPHREY

Over thousands of years, the scientific perspective upon the world has focused attention upon the simplicities and regularities of Nature. Those regularities have been found to reside in the rules governing the events that we see around us, rather than in the structure of the events themselves. The world is full of complex structures and erratic events that are the outcomes of a small number of simple and symmetrical laws. As we have learned, this is possible because the outcomes of the laws of Nature need not possess the symmetrical properties of the laws themselves. Laws can be the same everywhere and at all times; their outcomes need not be. This is how the Universe spawns complexity from simplicity. It is why we can talk of finding a Theory of Everything, yet fail to understand a snowflake.

Until quite recently, sciences like physics emphasized the deduction and confirmation of the laws and regularities of the world. The teaching of science was built around the simple, soluble problems that could be dealt with using pencil and paper. Since the early 1980s, there has been a change. The availability of small, inexpensive, and powerful computers with good interactive graphics has enabled large, complex, and disordered situations to be studied observationally. Experimental mathematics has been invented. The computer can be programmed to simulate the evolution of complicated systems, and their long-term behaviour studied, modified, and replayed. One can even construct virtual realities obeying laws of Nature that are not our own, and simply explore. By these means, the study of chaos and complexity has become a subculture within science. The study of the simple, exactly soluble problems of science has been augmented by a growing appreciation of the vast complexity that has to be expected in situations where many competing influences are at work. Prime candidates are supplied by systems that evolve in their environments by natural selection and, in so doing, modify those environments in complicated ways. One early discovery that emerged from these studies was of the ubiquity of chaotic behaviour—that is, behaviour displaying delicate sensitivity to small changes, so that any ignorance of its current state leads to complete ignorance about its state after a short period of time. Weather prediction suffers from this problem. We are poor at predicting tomorrow's weather because of our ignorance of the state of today's weather, not because we do not know how weather systems change. As our appreciation for the nuances of chaotic behaviour has matured by exposure to natural examples, novelties have emerged. Chaos and order have been found to coexist in a curious symbiosis. Imagine a very large egg-timer in which sand is falling, grain by grain, to create a growing pile of sand. The pile amasses in an erratic manner. Sand-falls of all sizes occur, and their effect is to maintain the overall gradient of the pile of sand in equilibrium, just on the verge of collapse. This self-sustaining process has been dubbed 'self-organizing criticality', by its discoverer Per Bak. At a microscopic level the process is chaotic. If there is nothing peculiar about the sand, such as to render avalanches of one size more or less probable than others, then the frequency with which avalanches occur is proportional to some mathematical power of their size. (The avalanches are 'scale-free' processes, just like the noise sources we considered in the last chapter when discussing the pattern of musical sounds.) There are many natural systems, such as earthquakes, and man-made ones, such as stock-market crashes, in which a concatenation of local processes combine to maintain a semblance of equilibrium in this way. Order develops on a large scale through the combination of many chaotic small-scale events that hover on the brink of instability. The pile of sand is always critically poised, and the next avalanche could be of any size; yet its effect is to maintain a well-defined overall slope of sand. The course of life on Earth might well be described by such a picture. The chain of living creatures maintains an overall balance dictated by the second law of thermodynamics, as we saw in Chapter 3, despite the constant impact of extinctions, changes of habitat, disease, and disaster, that conspire to create local 'avalanches'. Occasional extinctions open up new niches, and allow diversity to flourish anew, until equilibrium is temporarily re-established. A picture of the living world poised in a critical state, in which local chaos sustains global stability, is Nature's subtlest compromise. Complex adaptive systems thrive in the hinterland between the inflexibilities of determinism and the vagaries of chaos. There, they get the best of both worlds: out of chaos springs a wealth of alternatives for natural selection to sift, while the rudder of determinism sets a clear overall course.

Stephen Wolfram has argued that there exists a 'Principle of Computational Equivalence' in the Universe to the effect that natural processes and structures as well as humanly generated ones all have a similar level of complexity when viewed as computational processes. In some sense they are all as complicated as they can be. The brain and the rules of arithmetic may appear to be very different in terms of their complexity, but at root they have the same capabilities and the same unpredictabilities. This may offer some new insight into the remarkable usefulness of mathematics as a description of the world. Ultimately this is a reflection of the abundance of simple computable operations. It is not a mystery that the Universe is mathematical, but it is mysterious that such simple mathematics is so far reaching and tells us so much about the workings of the Universe. There also seems to be no reason why our mental capabilities are able to grasp so much of the deep structure of the Universe. Mere survival didn't need that. Computational equivalence suggest that there is a match between our minds and the environment around us that reflects the universality of computation in thought and natural processes. This tentative idea resonates strongly with one of the principal messages of this book. The propensities and sensitivities of the human mind have been inherited from the complexity of the environment in which it has evolved and acclimatized. We reflect many of the features of the laws of Nature and the structure of the Universe around us.

We have introduced these ideas to highlight a change of scientific perspective on the world. Science had for long emphasized the regularities and commonalities behind the appearances. The search for 'laws', 'invariances', 'constants', 'equations', 'solutions', 'periodicities', and 'principles'—this is the stuff of classical science. Pattern was the thing. Collecting butterflies and plants, listing all the stars in the sky: these are all very well. But they are not scientific activities until they seek to make sense of what they find, and sift sense from nonsense by predicting what they should find in the future. This search for simplicity and order, under the assumption of common laws that link the present to the future and the past, has directed the development of science during the past three hundred years. But complexity is not so simple. Only with the coming of studies of the complex, by means of new technologies, has the scientific eye turned to the problem of explaining diversity, asymmetry, and irregularity.

If we turn from the scientific to the artistic perspective on the world, we find an interesting contrast. Where science has progressed by searching for commonalities and patterns, the arts have celebrated diversity and have resisted attempts to encapsulate their activities in rules and formulae. They are the ultimate manifestations of the unpredictabilities and asymmetries of Nature. After all, what more chaotically unpredictable outcomes are there than some of those that issue from the human mind? So intractable has been the problem of finding pattern in creative activity that few would even seek it. If one looks not at science and art, but at scientists and artists, one finds a reflection of this divide. Two populations that overlap only a little, convergent thinkers and divergent thinkers, specialists and generalists—these labels reflect the differences of which we speak.

We are left to draw one last lesson. While science has enlarged its past horizons beyond order and symmetry to embrace diversity and unpredictability, the humanities have yet to appreciate the full force of commonality and pattern as a unifying factor in the interpretation of human creativity. Just as science has begun to appreciate that its view of Nature must reconcile simplicity and complexity, so the arts and humanities must appreciate the lessons to be drawn from the regularities of Nature. It is not enough to collect examples of diversity: the coexistence of diversity with universal behaviour is what requires exploration and reconciliation.

No mind was ever a *tabula rasa.* We enter the world with minds that possess an innate ability to learn. What we learn; how we learn; what we notice; and what we know but never learned—these things bear witness to our past in subtle ways. Creativity is not as untrammelled as it seems. Our humanity derives from shared experiences in the remote past, when many of our instincts and propensities were acquired as adaptations to a universal environment that set our ancestors common problems to overcome. Our minds developed susceptibilities that aided the solution of those problems. Many of those problems are no longer evident; hence some of our senses and sensibilities are adaptations to situations that no longer challenge us. They may even handicap us. While we can overwrite these inherited responses with learning, they remain (sometimes unnoticed) to spark our emotions in the absence of experience. Sometimes, even the overlay of conscious rationality fails to displace these inbred instincts. The sudden appearance of a flower, or the view from a great height: these experiences conjure up latent instincts, laid down and perpetuated in survivors over millions of years.

We are not all-purpose problem-solvers. The history of the human race has selected for the development of specific forms of analysis and response. Many features of our environment, in the widest sense of the world, have become internalized in our mental picture of the world. Our responses to those features have been sifted by natural selection. We sometimes respond to indicators, or symbols, that provide only partial cues about a potentially vital aspect of the environment. In this book we have looked at some of the ways in which the Universe has imposed aspects of its structure upon us by the inevitability of the forces of Nature, and we have considered the need for living things to adapt to their environments. In a world where adapters succeed, but non-adapters fail, one expects to find vestigial remnants of adaptations that once served other primary purposes. Many of these adaptations are subtle, and they have given rise to a suite of curious byproducts, some of which have played a role in determining our aesthetic sense. We are products of a past world where sensitivities to certain things were a matter of life and death.

In the past, the humanities and the sciences of human behaviour have been dominated by their celebration of the diversity of human behaviour.

Anthropologists were delighted to find new customs, novel habits, and different practices the world over. Common factors were ignored as uninteresting. Sometimes it became a little too easy to find what one was looking for. As any skilled cross-examiner knows, it is easy to find the truth one wants; not by inventing it, but by allowing the emergence of only that part of the whole truth that one wants to hear. A perspective that saw context, culture, and learning as the sole determinant of human behaviour thus made unquestioned progress. In contrast, we have in this book focused on the common factors of human experience. We believe that they are potentially of greater importance than the differences and, as scientists found long ago, much the easier to study. They link us to the universalities of the ancient environments in which the evolution of life occurred over enormous periods of time, long before the advent of civilization and recorded history; and they link us ultimately to the structure and origin of the Universe. The study of human actions, human minds, and human creativity has been quick to see complexity; slow to appreciate simplicity. Science, quick to see uniformity, has at last begun to appreciate diversity; but there is much for the creative arts to learn from the unity of the Universe about the propensities of our senses and the sights and sounds that excite them. And science, in its turn, will discover much about the emergence of complex organized structures from a renewed study of the mind's most artful inventions: a place where two ways meet.

Bibliography

Now, I realized that not infrequently books speak of books: it is as if they spoke among themselves. In the light of this reflection, the library seemed all the more disturbing to me. It was then the place of long, centuries-old murmuring, an imperceptible dialogue between one parchment and another, a living thing, a receptacle of powers not to be ruled by a human mind, a treasure of secrets emanated by many minds, surviving the death of those who had produced them or had been their conveyors.

UMBERTO ECO

Chapter 1: Tales of the unexpected

Appleyard, B. (1992) *Understanding the Present: Science and the Soul of Modern Man.* Doubleday, New York.

Dissanayake, E. (1988) *What is Art For?* University of Washington Press, Seattle.

Kuhn, T. (1962) *The Structure of Scientific Revolutions.* University of Chicago Press.

Shklovskii, I. S., and Sagan, C. (1967) *Intelligent Life in the Universe.* Dell, New York.

Chapter 2: The impact of evolution

■ A room with a view: matters of perspective

Alpers, S. (1983) *The Art of Describing: Dutch Art in the Seventeenth Century.* University of Chicago Press.

Arnheim, R. (1954) *Art and Visual Perception.* University of California Press, Berkeley.

Gardner, H. (1982) *Art, Mind and Brain.* Basic Books, New York.

Gombrich, E. H. (1979a) *Art and Illusion.* Phaidon, London.

Gombrich, E. H. (1979b) *The Sense of Order.* Phaidon, London.

Homer, W. I. (1964) *Seurat and the Science of Painting.* MIT Press, Massachusetts.

Humphrey, N. (1973) The illusion of beauty. *Perception* **2**, 429–39.

Ivins, W. (1946) *Art and Geometry: A Study in Space Intuitions.* Dover Books, New York.

Kandinsky, W. (1979) *Point and Line to Plane.* Dover Books, New York.

Richardson, J. A. (1971) *Modern Art and Scientific Thought.* University of Illinois Press, Urbana.

Shalin, L. (1991) *Art and Physics: Parallel Visions in Space, Time and Vision*. Morrow, New York.

Yee, C. (1938) *The Silent Traveller*. Country Life, London.

■ The mind-benders: distortions of thought and space

Barlow, H. (1961) The coding of sensory messages. In *Current Problems in Animal Behaviour* (ed. W. Thorpe and O. Zangwill), pp. 331–60. Cambridge University Press.

Barrett, C. (1971) *An Introduction to Optical Art*. Studio Vista, London.

Brush, S. (1987) The nebular hypothesis and the evolutionary worldview. *History of Science* **25**, 246–78.

Carraher, R. G., and Thurston, J. B. (1966) *Optical Illusions and the Visual Arts*. Reinhold/Studio Vista, New York.

Ernst, B. (1986) *The Eye Beguiled: Optical Illusions*. Taschen, Cologne.

James, W. (1892) *The Principles of Psychology*. Macmillan, London.

Kant, I. (1781) *The Critique of Pure Reason*. In *Encyclopedia Britannica Great Books*. Encyclopedia Britannica Inc., Chicago.

MacKay, D. M. (1957) Moving visual images produced by regular stationary patterns. *Nature* **180**, 849–50.

Necker, L. (1832) Observations on some remarkable phenomena seen in Switzerland: and on an optical phenomenon which occurs when viewing a figure of a crystal or geometrical solid. *London and Edinburgh Philosophical Magazine and Journal of Science* **1**, 329–37.

Richards, J. (1979) The reception of a mathematical theory: non-Euclidean geometry in England 1868–1883. In *Natural Order: Historical Studies of Scientific Culture* (ed. B. Barnes and S. Shapin). Sage, Beverly Hills.

Richards, R. J. (1987) *Darwin and the Emergence of Evolutionary Theories of Mind and Behavior*. University of Chicago Press.

Vitz, P. C., and Glimcher, A. B. (1984) *Modern Art and Modern Science*. Praeger, New York.

Waddington, C. H. (1969) *Behind Appearance: A Study of the Relations between Painting and the Natural Sciences in this Century*. Edinburgh University Press.

Wade, N. (1978) Op art and visual perception. *Perception* **7**, 21–46.

■ The inheritors: adaptation and evolution

Bonnor, J. T. (1988) *The Evolution of Complexity by Means of Natural Selection*. Princeton University Press, New Jersey.

Dawkins, R. (1976) *The Selfish Gene*. Oxford University Press.

Dawkins, R. (1983) Universal Darwinism. In *Evolution from Molecules to Men* (ed. D. S. Bendall), pp. 403–25. Oxford University Press.

Dawkins, R. (1986) *The Blind Watchmaker*. Longman, London.

Gould, S. J., and Lewontin, R. (1979) The spandrels of San Marco and the panglossian paradigm: a critique of the adaptationist programme. *Proceedings of the Royal Society* **B205**, 581–98.

Kauffman, S. (1991) *The Origins of Order*. Oxford University Press.

Kimura, M. (1983) *The Neutral Theory of Molecular Evolution*. Cambridge University Press.

Lewontin, R. C. (1978) Adaptation. *Scientific American* (Sept.), pp. 212–30.

Lewontin, R. C. (1991) *The Doctrine of DNA*. Penguin Books, Harmondsworth.

Maynard Smith, J. (1989) *Evolutionary Genetics*. Oxford University Press.

Mayr, E. (1991) *One Long Argument: Charles Darwin and the Genesis of Modern Evolutionary Thought*. Harvard University Press, Massachusetts.

Penrose, R. (1989) *The Emperor's New Mind: Concerning Computers, Minds, and the Laws of Physics*. Oxford University Press.

Smith, C. U. M. (1976) *The Problem of Life: an Essay in the Origins of Biological Thought*. Macmillan, London.

Sober, E. (1981) The evolution of rationality. *Synthèse* **46**, 95–120.

Stenseth, N. C., and Maynard Smith, J. (1984) Coevolution in ecosystems: Red Queen evolution or stasis? *Evolution* **38**, 870–80.

Van Valen, L. (1973) A new evolutionary law. *Evolutionary Theory* **1**, 1–30.

Williams, G. C. (1966) *Adaptation and Natural Selection*. Prometheus, New York.

Wilson, E. O. (1975) *Sociobiology: The New Synthesis*. Harvard University Press, Massachusetts.

After Babel: a linguistic digression

Bickerton, D. (1990) *Language and Species*. University of Chicago Press.

Chomsky, N. (1972) *Language and Mind* (rev. edn). Harcourt, Brace, Jovanovich, New York.

Chomsky, N. (1975) *Reflections on Language*. Pantheon, New York.

Chomsky, N. (1980) *Rules and Representations*. Columbia University Press, New York.

Chomsky, N. (1986) *Knowledge of Language: Its Nature, Origin, and Use*. Praeger, New York.

Chomsky, N. (1991) Linguistics and cognitive science: problems and mysteries. In *The Chomskyan Turn* (ed. A. Kasher), pp. 26–53. Blackwell, Cambridge, Massachusetts.

Flavell, P. (1963) *The Developmental Psychology of Jean Piaget*. Van Nostrand, Princeton.

Fry, R. (1920) *Vision and Design*. Meridian, New York.

Lennenberg, E. H. (1967) *Biological Foundations of Language*. Wiley, New York.

Lieberman, P. On the evolution of language: a unified view. *Cognition* **2**, 59–95.

Pinker, S. (1994) *The Language Instinct: the New Science of Language and Mind*. Penguin Books, Harmondsworth.

Pinker, S., and Bloom, P. (1990) Natural language and natural selection. *Behavioural and Brain Sciences* **13**, 707–84.

Steiner, G. (1992) *After Babel* (2nd edn.). Oxford University Press.

A sense of reality: the evolution of mental pictures

Biederman, C. (1948) *Art as the Evolution of Visual Knowledge*. Red Wing, Minnesota.

Churchland, P. M. (1979) *Scientific Realism and the Plasticity of Mind*. Cambridge University Press.

Delbrück, M. (1986) *Mind From Matter*. Blackwell, Oxford.

Laughlin, C. D., and D'Aquila, E. G. (1974) *Biogenetic Structuralism*. Columbia University Press, New York.

Lorenz, K. (1962) Kant's doctrine of the *a priori* in the light of contemporary biology. In *Yearbook of the Society for General Systems Research*, Vol. VII, pp. 23–35. Society for General Systems Research, New York. [Also reprinted in R. I. Evans, *Konrad Lorenz: the Man and his Ideas*, pp. 181–217. Harcourt, Brace, Jovanovich, New York. Both are translations of the German original, first published in 1941.]

Lorenz, K. (1978) *Behind the Mirror*. Harcourt, Brace, Jovanovich, New York.

Munz, P. (1993) *Philosophical Darwinism*. Routledge, London.

O'Hear, A. (1987) Has the theory of evolution any relevance to philosophy? *Ratio* **29**, 16–35.

Plotkin, H. (1994) *The Nature of Knowledge: Concerning Adaptations, Instinct and the Evolution of Intelligence*. Penguin Books, Harmondsworth.

Quine, W. V. O. (1969) Epistemology naturalized. In *Ontological Relativity and Other Essays*. Columbia University Press, New York.

Quine, W. V. O. (1973) *Roots of Reference*. Open Court, Peru, Illinois.

Riedl, R. (1984) *Biology of Knowledge: The Evolutionary Basis of Reason*. Wiley, Chichester.

Von Schilcher, F., and Tennant, N. (1984) *Philosophy, Evolution and Human Nature*. Routledge and Kegan Paul, London.

Wuketits, F. M. (1990) *Evolutionary Epistemology and Its Implications for Humankind*. State University of New York Press.

■ The care and maintenance of a small planet: cosmic environmentalism

Alvarez, L. W., Alvarez, W., Asaro, F., and Michel, H. V. (1980) Extraterrestrial cause for the Cretaceous-Tertiary extinction. *Science* **208**, 1095–108.

Alvarez, L. W., Kauffman, E. G., Surlyk, F., Alvarez, W., Asaro, F., and Michel, H. V. (1984) Impact theory of mass extinctions and the invertebrate fossil record. *Science* **223**, 1135–41.

Barrow, J. D. (1994) *The Origin of the Universe*. Basic Books, New York.

Barrow, J. D., and Silk, J. (1993) *The Left Hand of Creation: the Origin and Evolution of the Universe* (2nd edn.). Oxford University Press. [First edition published by Basic Books, New York (1983).]

Piel, G. (1992) *Only One World: Our Own to Make and to Keep*. Freeman, New York.

Raup, D. M. (1986) Biological extinction in Earth history. *Science* **21**, 1528–33.

Raup, D. M. (1987) Mass extinction: a commentary. *Paleontology* **30**, 1–13.

Sepkowski, J. J. (1988) Extinctions of life. *Los Alamos Science* **16**, 36–49.

Stanley, S. M. (1987) *Extinction*. Freeman, New York.

■ Gravity's rainbow: the fabric of the world

Barrow, J. D. (1988) *The World within the World*. Clarendon Press, Oxford.

Barrow, J. D. (1991) *Theories of Everything: the Quest for Ultimate Explanation*. Clarendon Press, Oxford.

Davies, P. C. W. (ed.) (1989) *The New Physics*. Cambridge University Press.

Dyson, F. (1971) Energy in the Universe. *Scientific American* (Sept.), 50–9.

Fritzsch, H. (1983) *Quarks*. Simon and Schuster, New York.

Hawking, S. W. (1988) *A Brief History of Time*. Bantam Books, New York.

Pagels, H. (1985) *Perfect Symmetry*. Michael Joseph, London.

Weinberg, S. (1983) *The Discovery of Subatomic Particles*. Freeman, New York.

Zee, A. (1986) *Fearful Symmetry: The Search for Beauty in Modern Physics*. Macmillan, New York.

▪ Chronicle of a death foretold: of death and immortality

Bell, G. (1984) Evolutionary and nonevolutionary theories of senescence. *American Naturalist* **124**, 600–3.

Crowe, M. J. (1986) *The Extraterrestrial Life Debate 1750–1900*. Cambridge University Press.

Davies, P. C. W. (1995) *Are We Alone?* Penguin, London.

Dick, S. J. (1982) *Plurality of Worlds: The Origins of the Extraterrestrial Life Debate from Democritus to Kant*. Cambridge University Press.

Drake, F. (1988) The search for extraterrestrial life. *Los Alamos Science* **16**, 50–69.

Fogg, M. J. (1987) Temporal aspects of the interaction among the first galactic civilizations: the 'Interdict Hypothesis'. *Icarus* **69**, 370–84.

Lewis, C. S. (1938) *Out of the Silent Planet*. John Lane, London.

Lightman, A. (1993) *Einstein's Dreams*. Bloomsbury, London.

Medawar, P. (1981) Old age and natural death. In *The Uniqueness of the Individual*. Dover Books, New York.

Williams, G. (1957) Pleiotropy, natural selection, and the evolution of senescence. *Evolution* **11**, 398–411.

▪ The human factor: light in the darkness

Bronowski, J. (1975) *Science and Human Values*. Harper and Row, New York.

Gould, S. J. (1985) Mind and supermind. In *The Flamingo's Smile, Reflections in Natural History*, pp. 392–402. Norton, New York.

Harrison, E. R. (1987) *Darkness at Night*. Harvard University Press, Massachusetts.

▪ The world is not enough: the grand illusion

Barrow, J. D. (1992) *Pi in the Sky*, Oxford University Press: Oxford.

Barrow, J. D. (1998) *Impossibility*, Jonathan Cape, London, chap. 8.

Barrow, J. D. (2002) *The Constants of Nature: From Alpha to Omega*, Jonathan Cape, London.

Barrow, J. D. (2003) Glitch!, *New Scientist* **178** No. 2398, 7 June, 44–45.

Bostrom, N. http://www.simulation-argument.com and *New Scientist*, 27 July 2002, 48.

Davies, P. C. W. A brief history of the multiverse, *New York Times*, 12 April, 2003.

Hanson, R. (2001) How to live in a simulation, *Journal of Evolution and Technology* 7 (http://www.transhumanist.com)

Harrison, E. R. (1995) The natural selection of universes containing intelligent life, *Quart. Jl. Roy. Astron. Soc.* **36**, 193.

MacKay, D. (1974) *The Clockwork Image*, IVP, London.

Popper, K. (1950) *British Journal for the Philosophy of Science* **1**, 117 and 173.

Webb, J. K., Murphy, M., Flambaum, V., Dzuba, V., Barrow, J. D., Churchill, C., Prochaska, J., and Wolfe, A. (2001). Further evidence for cosmological evolution of the fine structure constant, *Physical Review Letters* **87**, 091301.

Wolfram, S. (2002) *A New Kind of Science*, Wolfram Inc., Illinois.

Chapter 3: Size, life, and landscape

■ A delicate balance: equilibria in the Universe

Barrow, J. D., and Tipler, F. J. (1986) *The Anthropic Cosmological Principle*. Clarendon Press, Oxford.

Carr, B. J., and Rees, M. J. (1979) The Anthropic Principle and the structure of the physical world. *Nature* **278**, 605–12.

Davis, P. C. W. (1982) *Accidental Universe*. Cambridge University Press.

Press, W. H., and Lightman, A. (1983) Dependence of macrophysical phenomena on the values of the fundamental constants. *Proceedings of the Royal Society* **A310**, 323–36.

Weisskopf, V. F. (1975) Of atoms, mountains, and stars: a study in qualitative physics. *Science* **187**, 605–12.

■ Of mice and men: life on Earth

Alexander, R. McN. (1971) *Size and Shape*. Arnold, London.

Alexander, R. McN. (1989) *Dynamics of Dinosaurs and Other Extinct Giants*. Columbia University Press, New York.

Barrow, J. D. (1993) Sizing up the Universe. *Focus* (Oct.), 70–3.

Belloc, H. (1959) The Waterbeetle. In *Cautionary Verses* (illustrated by B. T. B. Bentley). Knopf, New York.

Bonner, J. T., and McMahon, T. A. (1983) *On Size and Life*. Freeman, New York.

Greenhill, A. G. (1881) Determination of the greatest height consistent with stability that a vertical pole or mast can be made, and of the greatest height to which a tree of given proportions can grow. *Proceedings of the Cambridge Philosophical Society* **4**, 65–73.

Hildebrandt, S., and Tromba, A. (1985) *Mathematics and Optimal Form*. Freeman, New York.

Lietzke, M. H. (1956) Relation between weight-lifting totals and body weight. *Science* **124**, 486–7.

Pennycuick, C. J. (1992) *Newton Rules Biology*. Oxford University Press, New York.

Thompson, D'A. W. (1917) *On Growth and Form*. Cambridge University Press.

Tóth, L. F. (1964) What the bees know and what they do not know. *Bulletin of the American Mathematical Society* **70**, 468–81.

The jagged edge: living fractals

Briggs, J. (1992) *Fractals: The Patterns of Chaos*. Simon and Schuster, New York.

Mandelbrot, B. (1977a) *Fractals: Form, Chance, and Dimension*. Freeman, New York.

Mandelbrot, B. (1977b) *The Fractal Geometry of Nature*. Freeman, New York.

Peitgen, H.-O., and Richter, P. H. (1986) *The Beauty of Fractals*. Springer, Berlin.

Bilateral agreements: appreciating curves

Barrow, J. D. (2003) Art and Science–Les Liaisons Dangereuses? In *Art and Complexity*, eds. J. Casti and A. Karlquist, Elsevier, Amsterdam, pp. 1–20.

Birkhoff, G. (1993) *Aesthetic Measure*, Harvard University Press, Cambridge.

Fractal Expressionism: the strange case of Jack the Dripper

Taylor, R. P., Micolich, A. P. and Jonas, D. (1999) Fractal Analysis of Pollock's Drip Paintings, *Nature* **399**, 422; (1999) Fractal Expressionism, *Physics World* **12**, 25; (2003) The Construction of Pollock's Fractal Drip Paintings, *Leonardo* **35**, 203.

War and peace: size and culture

Bonner, J. T. (1965) *Size and Cycle*. Princeton University Press, New Jersey.

Bonner, J. T. (1969) *The Scale of Nature*. Harper and Row, New York.

Bonner, J. T. (1980) *The Evolution of Culture in Animals*. Princeton University Press, New Jersey.

Bortz, W. M. (1985) Physical exercise as an evolutionary force. *Journal of Human Evolution* **14**, 145–55.

Calder, W. A. III (1984) *Size, Function, and Life History*. Harvard University Press, Massachusetts.

Haldane, J. B. S. (1928) On being the right size. In *Possible Worlds*, pp. 20–8. Harper, New York.

McNab, B. K. (1983) Energetics, body size, and the limits to endothermy. *Journal of Zoology (London)* **199**, 1–29.

Moog, F. (1948) Gulliver was a bad biologist. *Scientific American* (Nov.), pp. 52–5.

Pilbeam, D., and Gould, S. J. (1974) Size and scaling in human evolution. *Science* **186**, 892–901.

Swift, J. (1726) *Gulliver's Travels*. Reprint of 5th edn. (1710), Oxford University Press, London, 1919.

Went, F. W. (1968) The size of man. *American Scientist* **56**, 400–13.

Far from the madding crowd: the size of populations

Brown, J. H., and Maurer, B. A. (1989) Macroecology: the division of food and space among species on continents. *Science* **243**, 1145–50.

Cerling, T. E. (1991) Carbon dioxide in the atmosphere: evidence from Cenozoic and Mesozoic paleosols. *American Journal of Science* **291**, 377–400.

Colinvaux, P. (1978) *Why Big Fierce Animals are Rare: an Ecologist's Perspective*. Princeton University Press, New Jersey.

Damuth, J. (1981) Population density and body size in mammals. *Nature* **290**, 699–700.

Elton, C. (1927) *Animal Ecology*. Macmillan, New York.

Farlow, J. O. (1987) Speculations about the diet and digestive physiology of herbivorous dinosaurs. *Paleobiology* **13**, 60–72.

Farlow, J. O. (1993) On the rareness of big, fierce animals: speculations about the body sizes, population densities, and geographic ranges of predatory mammals and large carnivorous dinosaurs. *American Journal of Science* **293A**, 167–99.

Janis, C. M., and Carrano, M. (1994) Scaling of reproductive turnover in archosaurs and mammals: why are large terrestrial mammals so rare? In Kurten Memorial Volume (ed. B. M. Fortelius, L. Werdelin, and A. Forsten), *Acta Zoologica Fennica*.

May, R. (1984) In *Diversity of Insect Faunas*, pp. 188–204 (ed. L. A. Mound and N. Waloff). Blackwell Scientific Publications, Oxford.

Peters, R. H. (1983) *The Ecological Implications of Body Size*. Cambridge University Press.

Randall, J. E. (1975) Size of the Great White Shark (*Carcharodon*). *Science* **181**, 169–70.

Spotila, J. R., O'Connor, M. P., Dodson, P., and Paladino, F. V. (1991) Hot and cold running dinosaurs: body size, metabolism and migration. *Modern Geology* **16**, 203–27.

Van Valkenburgh, B. (1988) Trophic diversity in past and present guilds of large predatory mammals. *Paleobiology* **14**, 155–73.

■ *Les liaisons dangereuses*: complexity, mobility, and cultural evolution

Bonner, J. T. (1988) *The Evolution of Complexity*. Princeton University Press, New Jersey.

Bonner, J. T. (1989) *The Evolution of Culture in Animals*. Princeton University Press, New Jersey.

DeBrul, E. L. (1962) The general phenomenon of bipedalism. *American Zoologist* **2**, 205–8.

Hebb, D. O. (1955) The mammal and his environment. *American Journal of Psychology* **3**, 826–31.

Hewes, G. W. (1961) Food transportation and the origins of bipedalism. *American Anthropologist* **63**, 687–710.

Konner, M. (1982) *The Tangled Wing: Biological Constraints on the Human Spirit*. Penguin Books, Harmondsworth.

Tooby, J., and Cosmides, L. (1990) The past explains the present: emotional adaptations and the structure of ancestral environments. *Ethology and Sociobiology* **11**, 375–424.

■ Network news: branching out

Banavar, J. R., Maritan, A. and Rinaldo, A. (1999) Size and form in efficient transport networks, *Nature* **399**, 130–2.

Banavar, J. R., Damuth, J., Maritan A. and Rinaldo, A. (2002) Supply-demand balance and metabolic scaling, *Proceedings of the National Academy of Sciences* **99**, 10506–9.

Schmidt-Nielsen, K. (1984) *Why is Animal Size so Important?*, Cambridge University Press.

West, G. B., Brown, J. H. and Enquist, B. J. (1997) A general model for the origin of allometric scaling laws in biology, *Science* **276**, 122–6.

West, G. B., Brown, J. H. and Enquist, B. J. (1999a) The fourth dimension and life: fractal geometry and allometric scaling of organisms, *Science* **284**, 1677–9.

West, G. B., Brown, J. H. and Enquist, B. J. (1999b), A general model for the structure of plant vascular systems, *Nature* **400**, 644–7.

▦ The Go-Betweenies: messing with Mister In-Between

Brace, C. L. (1964) The fate of the Neanderthals: a consideration of hominid catastrophism. *Current Anthropology* **5**, 3–43.

Clutton-Brock, T. H., and Harvey, P. (1981) Primate home range size and metabolic needs. *Behavioural Ecology and Sociobiology* **8**, 151–5.

Foley, R. (1987) *Another Unique Species*. Longman, London.

Isaac, G. (1971) The diet of early Man: aspects of archaeological evidence from Lower and Middle Pleistocene sites in Africa. *World Archaeology* **2**, 278–98.

Isaac, G., and Crader, D. (1981) To what extent were early hominids carnivorous? An archaeological perspective. In *Omnivorous Primates* (ed. R. S. Harding and G. Teleki), pp. 37–103. Columbia University Press, New York.

Jerison, H. J. (1973) *Evolution of the Brain and Intelligence*. Academic Press, New York.

McHenry, M. (1982) The pattern of human evolution: studies on bipedalism, mastication and encephalization. *Annual Review of Anthropology* **11**, 151–73.

Newman, R. W. (1970) Why man is a sweaty and thirsty naked animal: a speculative review. *Human Biology* **42**, 12–27.

Pfeiffer, J. E. (1982) *The Emergence of Culture*. Harper, New York.

▦ The rivals: the evolution of cooperation

Axelrod, R. (1984) *The Evolution of Cooperation*. Penguin Books, Harmondsworth.

Axelrod, R., and Hamilton, W. D. (1981) The evolution of cooperation. *Science* **211**, 1390–6.

Hamilton, W. D. (1964) The genetical evolution of social behaviour. *Journal of Theoretical Biology* **7**, 1–52.

Lieberman, P. (1991) *Uniquely Human: the Evolution of Speech, Thought and Selfless Behaviour*. Harvard University Press, Massachusetts.

Maynard Smith, J. (1978) The evolution of behaviour. *Scientific American* (Sept.), pp. 176–93.

Maynard Smith, J. (1982) *Evolution and the Theory of Games*. Cambridge University Press.

Tooby, J., and Cosmides, L. (1992) Cognitive adaptations for social exchange. In *The Adapted Mind* (ed. J. H. Barkow, L. Cosmides, and J. Tooby), pp. 163–228. Oxford University Press, New York.

Trivers, R. L. (1971) The evolution of reciprocal altruism. *Quarterly Review of Biology* **46**, 35–57.

Trivers, R. L. (1981) *Social Evolution*. Benjamin-Cummings, New York.

▦ The secret garden: the art of landscape

Appleton, J. (1975) *The Experience of Landscape*. Wiley, New York.

Appleton, J. (ed.) (1980) *The Aesthetics of Landscape*. Rural Planning Services, Didcot, Oxfordshire.

Appleton, J. (1984) Prospects and refuges revisited. *The Landscape Journal* **3**, 91–103.

Appleton, J. (1990) *The Symbolism of Habitat.* University of Washington Press, Seattle.

Balling, J. D., and Falk, J. H. (1982) Development of visual preference for natural environments. *Environment and Behaviour* **14**, 5–28.

Hilden, O. (1965) Habitat selection in birds. *Annales Zoologici Fennici* **2**, 53–75.

Jenkins, I. (1958) *Art and the Human Enterprise.* Harvard University Press, Massachusetts.

Orians, G. H. (1980) Habitat selection: general theory and applications to human behaviour. In *Evolution of Human Social Behaviour* (ed. J. Lockard), pp. 49–66. Elsevier, New York.

Orians, G. H. (1986) An ecological and evolutionary approach to landscape aesthetics. In *Landscape Meanings and Values* (ed. E. Penning-Rowsell and D. Lowenthal), pp. 3–35, Allen and Unwin, London.

Orians, G. H., and Heerwagen, J. H. (1992) In *The Adapted Mind* (ed. J. H. Barkow, L. Cosmides, and J. Tooby), pp. 555–79. Oxford University Press, New York.

Tuan, Y. (1979) *Landscapes of Fear.* Pantheon, New York.

Ulrich, R. (1979) Visual landscapes and psychological well-being. *Landscape Research* **4**, 17–23.

Wilson, E. O. (1984) *Biophilia.* Harvard University Press, Massachusetts.

■ Figures in a landscape: the dilemma of computer art

Ascher, M. (1991) *Ethnomathematics.* Brooks/Cole, Pacific Grove, California.

Audsley, W. (1968) *Designs and Patterns from Historic Ornament.* Dover Books, New York.

Birkhoff, G. D. (1933) *Aesthetic Measure.* Harvard University Press, Massachusetts.

Burke, E. (1757) *A Philosophical Enquiry into the Origin of Our Ideas of the Sublime and Beautiful.* Revised edition (ed. J. T. Boulton) Blackwell, Oxford (1987).

Coxeter, H. S. M., Emmer, M., Penrose, R., and Teuber, M. L. (1986) *M. C. Escher: Art and Science.* Elsevier, Amsterdam.

Critchlow, K. (1984) *Islamic Patterns: An Analytical Approach.* Thames and Hudson, London.

Entsminger, G. (1989) Stochastic fiction: fiction from fractals. *Micro Cornucopia* (Sept./ Oct.), **49**, 96.

Escher, M. C. (1961) *The Graphic Work of M. C. Escher Introduced and Explained by the Artist.* Pan Books, London.

Field, M., and Golubitsky, M. (1991) *Symmetry in Chaos.* Oxford University Press.

Franke, H. W. (1971) *Computer Graphics, Computer Art.* Phaidon, London.

Grünbaum, B., and Shephard, G. C. (1987) *Tilings and Patterns.* Freeman, New York.

Kappraff, J. (1991) *Connections: the Geometric Bridge Between Art and Science.* McGraw-Hill, New York.

Latham, W., and Todd, S. (1992) *Evolutionary Art.* Academic Press, New York.

Mueller, R. E. (1967) *The Science of Art: The Cybernetics of Creative Communication.* Rapp and Whiting, London.

Mueller, R. E. (1972) Idols of computer art. *Art in America* (May/June), pp. 68–73.

Mueller, R. (1983) When is computer art art? *Creative Computing* (January), pp. 136–44.

Schaaf, W. L. (1951) Art and mathematics: a brief guide to source materials. *American Mathematical Monthly* **58**, 167–77.

Stevens, P. (1981) *A Handbook of Regular Patterns: An Introduction to Symmetry in Two Dimensions.* MIT Press, Massachusetts.

Vanderplas, J. M., and Garvin, E. A. (1959) The association value of random shapes. *Journal of Experimental Psychology* **57**, 147–54.

Vitz, P. C. (1966) Preference for different amounts of visual complexity. *Behavioural Science* **2**, 105–14.

Washburn, D. K., and Crowe, D. W. (1989) *Symmetries of Culture: Theory and Practice of Plane Pattern Analysis.* University of Washington Press, Seattle.

Midnight's children: a first glimpse of the stars

Aveni, A. (1980) *Skywatchers of Ancient Mexico.* University of Texas Press, Austin.

Santayana, G. (1955) *The Sense of Beauty.* Dover Books, New York (originally published 1896).

Chapter 4: The heavens and the Earth

The remains of the day: rhythms of life

Aveni, A. (1990) *Empires of Time: Calendars, Clocks, and Cultures.* Tauris, London.

Brown, F. (1962) *Biological Clocks.* Heath, Englewood, New Jersey.

Cazenave, A. (ed.) (1986) *Earth Rotation: Solved and Unsolved Problems.* Reidel, Dordrecht.

Saunders, D. (1977) *An Introduction to Biological Rhythms.* Wiley, New York.

Young, M. (1988) *The Metronomic Society: Natural Rhythms and Human Timetables.* Harvard University Press, Massachusetts.

Winfree, A. T. (1987) *The Timing of Biological Clocks.* Freeman, New York.

Empire of the Sun: the reasons for the seasons

Farhi, E., and Guth, A. (1987) An obstacle to creating a universe in the laboratory. *Physics Letters* **B183**, 149–55.

Harrison, E. R. (1995) The natural selection of universes containing intelligent life. *Quarterly Journal of the Royal Astronomical Society* **36**, 193.

Kandel, R. S. (1980) *Earth and Cosmos.* Pergamon Press, Oxford.

Paley, W. (1802) *Natural Theology.* In *The Works of William Paley* (ed. R. Lynam). London (1825).

Shu, F. (1982) *The Physical Universe.* University Science Books: Mill Valley, California.

Wallace, A. R. (1903) *Man's Place in the Universe.* Chapman & Hall, London.

Wylie, F. E. (1979) *Tides and the Pull of the Moon.* Greene Press, Brattleboro, Vermont.

Extrasolar planets: a case of spatial prejudice

The *Extrasolar Planets Encyclopedia* is available online at http://www.obspm.fr/encycl/encycl.html

■ A handful of dust: the Earth below

Anderson, D. L. (1991) *Theory of the Earth*. Blackwell Scientific Publications, Oxford.

Barrow, J. D. (1987) Observational limits on the time-evolution of extra spatial dimensions. *Physical Review* **D35**, 1805–10.

Cook, A. (1979) Geophysics and the human condition. *Quarterly Journal of the Royal Astronomical Society* **20**, 229–40.

Diamond, J. (1991) *The Rise and Fall of the Third Chimpanzee*. Vintage Books, London.

Dott, R. H., and Batten, R. L. (1988) *Evolution of the Earth* (4th edn.). McGraw-Hill, New York.

Grossman, L., and Larimer, J. W. (1974) Early chemical history of the solar system. *Reviews of Geophysics and Space Physics* **12**, 71–101.

Maurette, M. (1976) The Oklo Reactor. *Annual Reviews of Nuclear and Particle Science* **26**, 319–50.

Mian, Z. (1993) Understanding why Earth is a planet with plate tectonics. *Quarterly Journal of the Royal Astronomical Society* **34**, 441–8.

Pollard, W. G. (1979) The prevalence of Earthlike planets. *American Scientist* **67**, 653–9.

Teske, R. (1993) Planetary resources for extraterrestrial technology are unlikely. *Quarterly Journal of the Royal Astronomical Society* **34**, 335–6.

■ Pebble in the sky: the Moon above

Hartmann, W. K. (1983) *Moon and Planets*. Wadsworth, Belmont.

Krupp, E. (1983) *Echoes of the Ancient Skies: the Astronomy of Lost Civilizations*. Harper and Row, New York.

Marshack, A. (1972) *The Roots of Civilization*. McGraw-Hill, New York.

Ringwood, A. E. (1979) *The Origin of the Earth and the Moon*. Springer, Berlin.

■ Darkness at noon: eclipses

Asimov, I. (1941) *Nightfall and Other Stories*. Doubleday, New York. [The story 'Nightfall' appeared first in the September 1941 issue of the science fiction magazine *Astounding*, which featured it on the cover.]

Berendzen, R., Hart, R., and Seeley, D. (1976) *Man Discovers the Galaxies*. Neale Watson Academic, New York.

Brewer, B. (1991) *Eclipse* (2nd edn.). Earth View, Seattle.

McCrea, W. H. (1972) Astronomer's luck. *Quarterly Journal of the Royal Astronomical Society* **13**, 506–19.

Rescher, N. (1985) Extraterrestrial science. In *Extraterrestrials, Science and Alien Intelligence* (ed. E. Regis), pp. 83–116. Cambridge University Press.

Will, C. M. (1981) *Theory and Experiment in Gravitation Physics*. Cambridge University Press.

Zirker, J. B. (1984) *Total Eclipse of the Sun*. Van Nostrand, New York.

Hamlet's mill: the wandering Pole Star

Aveni, A. (1981) Tropical archeoastronomy. *Science* **213**, 161–71.

Krupp, E. (1991) *Beyond the Blue Horizon*. Oxford University Press, New York.

Santillana, G. de, and H. von Dechend (1969) *Hamlet's Mill*. Gambit, New York.

Unsold, A. (1977) *The New Cosmos* (2nd edn., trans. R. C. Smith). Springer, New York.

Paper moon: controlling chaotic planets

Imbrie, J., and Imbrie, K. P. (1979) *Ice Ages: Solving the Mystery*. Harvard University Press, Massachusetts.

Lambeck, K. (1980) *The Earth's Variable Rotation: Geophysical Causes and Consequences*. Cambridge University Press.

Laskar, J., and Robutel, P. (1993) The chaotic obliquity of the planets. *Nature* **361**, 608–12.

Laskar, J., Joutel, F., and Boudin, F. (1993) Orbital, precessional, and insolation quantities for the Earth from −20 Myr to +10 Myr. *Astronomy and Astrophysics* **270**, 522–33.

Laskar, J., Joutel, F., and Boudin, F. (1993) Stabilization of the Earth's obliquity by the Moon. *Nature* **361**, 615–17.

Ward, W. R. (1982) *Icarus* **50**, 444–8.

Mars in your eyes: they came from outer space

Catling, D. Planet Mars in Popular Culture. http://briefme.com/archive.php/article/6738

Lowell, P. (1895) *Mars*. Full text online at http://www.wanderer.org/references/lowells/Mars/

Wells, H. G. (1898) *The War of the Worlds*. Full text online at http://www.fourmilab.ch/etexts/www/warworlds/warw.html

The man who was Thursday: the origins of the week

Achelis, E. (1955) *Of Time and the Calendar*. Hermitage, New York.

Anon. (1931) The continuous working week in Soviet Russia. *International Labor Review* **23**, 157–80.

Burnaby, S. B. (1901) *Elements of the Jewish and Mohammedan Calendars*. George Bell, London.

Chamberlain, W. H. (1931) *The Soviet Planned Economic Order*. World Peace Foundation, Boston.

Colson, F. H. (1926) *The Week*. Cambridge University Press.

Cotton, P. (1933) *From Sabbath to Sunday*. Times Books, Bethlehem, Pennsylvania.

Cumont, F. (1960) *Astrology and Religion Among the Greeks and Romans*. Dover Books, New York.

Dio Cassius (1905) *Dio's Roman History*. Pafraets, New York.

Duran, D. (1971) *Book of the Gods and Rites and the Ancient Calendar*. University of Oklahoma Press, Norman.

Fagerlund, V. G., and Smith, R. H. T. (1970) A preliminary map of market periodicities in Ghana. *Journal of Developing Areas* **7**, 333–47.

Fleet, J. F. (1912) The use of the planetary names of the days of the week in India. *Journal of the Royal Asiatic Society* (new series) **44**, 1039–46.

Gandz, S. (1948) The origin of the planetary week or the planetary week in Hebrew literature. *Proceedings of the American Academy for Jewish Research* **18**, 213–54.

Gruliow, L. (1953) Significant Russian approval. *Journal of Calendar Reform* **23**, 101–5.

Hare, J. C. (1832) On the names of the days of the week. *The Philological Museum* **1**, 1–73.

Long, C. H. (1963) *Alpha: The Myths of Creation*. George Braziller, New York.

Meek, T. J. (1914) The Sabbath in the Old Testament. *Journal of Biblical Literature* **33**, 201–12.

Neugabauer, O. (1962) *The Exact Sciences in Antiquity*. Harper, New York.

Parry, A. (1940) The Soviet calendar. *Journal of Calendar Reform* **10**, 63–9.

Sachs, A. (1952) Babylonian horoscopes. *Journal of Cuneiform Studies* **6**, 49–75.

Strutynski, U. (1975) Germanic divinities in weekday names. *Journal of Indo-European Studies* **3**, 363–84.

Sydenham, M. J. (1974) *The First French Republic, 1792–1804*. Batsford, London.

Thomas, N. W. (1924) The week in West Africa. *Journal of the Royal Anthropological Institute* **54**, 183–209.

Van der Waerden, B. L. (1968) The date of invention of Babylonian planetary theory. *Archive for History of Exact Sciences* **5**, 70–8.

Zerubavel, E. (1977) The French Republican calendar: a case study in the sociology of time. *American Sociological Review* **42**, 868–76.

Zerubavel, E. (1985) *The Seven Day Circle: The History and Meaning of the Week*. Free Press, New York.

▓ Long day's journey into night: the origin of the constellations

Bauval, R., and Gilbert, A. (1994) *The Orion Mystery*. Heinemann, London.

Brown, R. (1899/1900) *Researches into the Origin of the Primitive Constellations of the Greeks, Phoenicians and Babylonians* (2 vols.). Williams and Norgate, London.

Crommelin, A. C. D. (1923) The ancient constellation figures. In *Splendour of the Heavens*. vol. 2 (ed. T. E. R. Phillips and W. H. Steavenson). Hutchinson, London.

Gurstein, A. (1993) On the origin of the zodiacal constellations, *Vistas in Astronomy*, **36**, 171–90.

Gurstein, A. (1994) Dating the origin of the constellations by precession, *Soviet Physics Doklady*, **39**, 575–8.

Gurstein, A. (1995a) Prehistory of zodiac dating: three strata of Upper Paleolithic constellations, *Vistas in Astronomy*, **39**, 347–62.

Gurstein, A. (1995b) When the zodiac climbed into the sky, *Sky and Telescope*, October, 28–33.

Gurstein, A. (1996) The great pyramids of Egypt as sanctuaries commemorating the origin of the zodiac: an analysis of astronomical evidence, *Soviet Physics Doklady* **41**, 228–32.

Gurstein, A. (1997a) In search of the first constellations, *Sky and Telescope*, June, 46–50.

Gurstein, A. (1997b) The Origins of the constellations, *American Scientist* **85**(3), 264–73 and 500–501.

Gurstein, A. (1997c) The evolution of the zodiac in the context of ancient oriental history, *Vistas in Astronomy* **41**, 507–25.

Higgins, R. (1973) *The Archaeology of Minoan Crete.* Bodley Head, London.

Krupp, E. C. (1994) Celestial reptiles. *Sky and Telescope* (April), 64–5.

Mair, G. R. (1921) *Aratus' Phaenomena* (trans.). Loeb Classical Library, Heinemann, London.

Maunder, E. W. (1909) *The Astronomy of the Bible.* Hodder and Stoughton, London.

Ovenden, M. W. (1966) The origin of the constellations. *Philosophical Journal* **3**, 1–18.

Roy, A. (1984) The origin of the constellations. *Vistas in Astronomy* **27**, 171–97.

Swartz, C. (1809) *Le Zodiaque expliquè*, 2nd edn. Migneret and Desenne, Paris, transl. from Swedish; 1st edn. 1807. This is catalogued as Item 5 in volume 47 of the *Bound Pamphlets* collection of the Royal Astronomical Society Library at Burlington House, London.

Study in scarlet: the sources of colour vision

Berlin, B., and Kay, P. (1969) *Basic Color Terms.* University of California Press, Berkeley.

Bohren, C. F. (1988) Understanding colors in Nature. *Pigment Cell Research* **1**, 214–22.

Borstein, M. H. (1974) The influence of visual perception on culture. *American Anthropologist* **77**, 774–98.

Boston, R. (1978) Black, white, and red all over: the riddle of color term salience. *Ethnology* **17**, 287–311.

Fox, H. M., and Vevers, G. (1960) *The Nature of Animal Colours.* Sedgwick and Jackson, London.

Gage, J. (1995), Colour and Culture, in Colour: Art & Science, eds. T. Lamb and J. Bourriau, Cambridge UP, pp. 175–193.

Gage, J. (1999) *Colour and Meaning: Art, Science and Symbolism*, Thames and Hudson, London.

Govardovskii, V. I. (1976) Comments on the sensitivity hypothesis. *Vision Research* **16**, 1363–4.

Humphrey, N. (1983a) *Consciousness Regained.* Oxford University Press.

Humphrey, N. (1983b) *A History of the Mind.* Vintage, London.

Jacobs, G. H. (1981) *Comparative Color Vision.* Academic Press, New York.

Kandinsky, W. (1977) *Concerning the Spiritual in Art* (trans. M. Sadler). Dover Books, New York. [First published 1912.]

Kay, P. (1975) Synchronic variability and diachronic change in basic colour terms. *Language in Society* **4**, 257–70.

Kay, P., and McDaniel, C. K. (1978) The linguistic significance of the meanings of basic color terms. *Language* **54**, 610–46.

Lyons, J. (1995) Colour in language. In *Colour: Art and Science* (eds. T. Lamb and J. Bourriau). Cambridge University Press, pp 194–224.

Lythgoe, J. N. (1979) *The Ecology of Vision.* Oxford University Press.

Ratliff, F. (1976) On the psychophysical base of color names. *Proceedings of the American Philological Society* **120**, 311–30.

Ray, V. F. (1952) Techniques and problems in the study of human color perception. *Journal of Anthropology* **8**, 251–9.

Roach, R. E., and Gordon, J. L. (1973) *The Light of the Night Sky*. Reidel, Dordrecht.

Sahlins, M. (1976) Colors and culture. *Semiotica* **16**, 1–22.

Shepard, R. N. (1992) In *The Adapted Mind: Evolutionary Psychology and the Generation of Culture* (ed. J. H. Barkow, L. Cosmides, and J. Tooby). Oxford University Press, New York.

Sobel, M. I. (1987) *Light*. University of Chicago Press.

Witkowski, S. R., and Brown, C. H. (1977) An explanation of color nomenclature universals. *American Anthropologist* **79**, 50–7.

Witkowski, S. R., and Brown, C. H. (1978) Lexical universals. *Annual Review of Anthropology* **7**, 427–51.

▪ Outward bound: the way of the world

Barrow, J. D. (2002) *The Constants of Nature: from alpha to omega*, Jonathan Cape, London.

Barrow, J. D. (2005), *The Infinite Book: a short guide to the boundless, timeless and endless*, Jonathan Cape, London.

Chapter 5: The natural history of noise

▪ The club of queer trades: soundscapes

Darwin, C. (1872) *The Expression of the Emotions in Man and Animals*. Murray, London.

Lomax, A. (1962) Song structure and social structure. *Ethnology* **1**, 425–51.

Needham, R. (1967) Percussion and transition. *Man* **2**, 606–14.

Roederer, J. G. (1984) The search for survival value in music. *Music Perception* **1**, 350–6.

Roederer, J. G. (1987) Why do we love music? A search for the survival value of music. In *Music in Medicine* (ed. R. Spintge and R. Droh). Springer, Heidelberg.

Sebeok, T. A. (1981) *The Play of Musement*, ch. 9. Indiana University Press, Bloomington.

Sloboda, J. A. (1985) *The Musical Mind: The Cognitive Psychology of Music*. Clarendon Press, Oxford.

Sloboda, J. A. (ed.) (1988) *Generative Processes in Music*. Clarendon Press, Oxford.

Storr, A. (1992) *Music and the Mind*. Harper Collins, New York.

Thorpe, W. H. (1966) Ritualization in the individual development of bird song. In J. Huxley (ed.) A discussion of ritualization of behaviour in animals and Man. *Philosophical Transactions of the Royal Society* **B772**, 351–8.

Thorpe, W. H. (1972) *Duetting and Antiphonal Song in Birds: Its Extent and Significance*. E. J. Brill, Leiden.

Tuzin, D. (1984) Miraculous voices: the auditory experience of numinous objects. *Current Anthropology* **25**, 579–96.

Von Frisch, K. (1974) *Animal Architecture*. Harcourt, Brace, Jovanovich, New York and London.

Sense and sensibility: a matter of timing

Edelman, G. (1992) *Bright Air, Brilliant Fire: On the Matter of the Mind.* Basic Books, New York.

Nettl, B. (1983) *The Study of Ethnomusicology: Twenty-nine Issues and Concepts.* University of Illinois, Urbana.

Turner, F., and Pöppel, E. (1983) The neural lyre: poetic meter, the brain, and time. *Poetry* **142**, 277–309.

Incidental music: a harmless by-product?

Cooke, D. (1959) *The Language of Music.* Oxford University Press.

Fry, R. (1924) *The Artist and Psychoanalysis.* Hogarth Press, London.

Gardner, H. (1983) *Frames of Mind.* Basic Books, New York.

Hanslick, E. (1957) *The Beautiful in Music* (trans. G. Cohen). Liberal Arts Press, Indianapolis.

Hesse, H. (1969) *The Glass Bead Game: Magister Ludi.* Holt, Rinehart and Winston, New York.

Meyer, L. B. (1956) *Emotion and Meaning in Music.* University of Chicago Press.

Tolstoy, L. (1960) *What is Art?* (trans. A. Mande). Liberal Arts Press, Indianapolis.

Weber, M. (1969) *The Rational and Social Foundations of Music.* Southern Illinois University Press, Carbondale.

The glass bead game: the music of the spheres

Buckert, W. (1972) *Lore and Science in Ancient Pythagoreanism.* Harvard University Press, Massachusetts.

Gorman, P. (1979) *Pythagoras: A Life.* Routledge, London.

Hunt, F. V. (1978) *Origins in Acoustics: The Science of Sound from Antiquity to the Age of Newton.* Yale University Press, New Haven.

James, J. (1993) *The Music of the Spheres—Music, Science and the Natural Order of the Universe.* Abacus, London.

Philip, J. A. (1966) *Pythagoras.* University of Toronto Press.

Schroeder, M. (1986) *Number Theory in Science and Communication* (2nd edn.). Springer, New York.

Zuckerkandl, V. (1956) *Sound and Symbol* (Vol. 2). Pantheon, New York.

Player piano: hearing by numbers

Barrow, J. D. (1992) *Pi in the Sky: Counting, Thinking, and Being.* Oxford University Press.

Butterworth, B. (1999) *The Mathematical Brain.* Macmillan, London.

Dehaene, S. (1997) *The Number Sense: How the Mind Creates Mathematics.* OUP, New York.

Dodge, C., and Jerse, T. A. (1985) *Computer Music.* Schirmer, New York.

Eigen, M., and Winkler, R. (1983) *Laws of the Game,* ch. 18. Penguin Books, London.

Ernest, P. (1998) *Social Constructivism as a Philosophy of Mathematics.* State University of New York.

Gardner, H. (1983) *Frames of Mind: The Theory of Multiple Intelligences*. Basic Books, New York.

Harwood, D. L. (1976) Universals in music: a perspective from cognitive psychology. *Ethnomusicology* **20**, 521–33.

Hersh, R. (1998) *What is Mathematics Really?* Vintage, London.

MacLane, S. (1986) *Mathematics: Form and Function*. Springer-Verlag, New York.

Maddy, P. (1992) *Realism in Mathematics*, Oxford University Press, New York.

Mathews, M. V., and Pierce, J. R. (eds.) (1990) *Current Directions in Computer Music Research*. MIT Press, Massachusetts.

Pickover, C. (1990) *Computers, Pattern, Chaos and Beauty*. Sutton, New York.

Pickover, C. (1992) *Mazes for the Mind*. St Martin's Press, New York.

Repp, B. (1990) Patterns of expressive timing in performances of a Beethoven minuet by 19 famous pianists. *Journal of the Acoustical Society of America* **88**, 622–41.

Road, C. (ed.) (1985) *Composers and the Computer*. Kaufman, California.

■ The sound of silence: decomposing music

Boden, M. (1990) *The Creative Mind: Myths and Mechanisms*. Weidenfeld and Nicholson, London.

Deutsch, D. (1984) Musical space. In *Cognitive Processes in the Perception of Art* (ed. W. R. Crozier and A. J. Chapman), pp. 253–87. Elsevier, New York.

Longuet-Higgins, H. C. (1962a) Letter to a musical friend. *Music Review* **23**, 244–8.

Longuet-Higgins, H. C. (1962b) Second letter to a musical friend. *Music Review* **23**, 271–80.

Longuet-Higgins, H. C. (1976) The perception of melodies. *Nature* **263**, 646–53.

Longuet-Higgins, H. C. (1978) The perception of music. *Interdisciplinary Science Reviews* **3**, 148–56.

Longuet-Higgins, H. C. (1994) Artificial intelligence and musical cognition. *Philosophical Transactions of the Royal Society London* **A349**, 103–13.

Steedman, M. (1994) The well-tempered computer. *Philosophical Transactions of the Royal Society London* **A349**, 115–31.

■ The imitation game: listlessness

Barrow, J. D. (1991) *Theories of Everything: The Quest for Ultimate Explanation*. Clarendon Press, Oxford.

Barrow, J. D. (1992) *Pi in the Sky: Counting, Thinking, and Being*. Clarendon Press, Oxford.

Chaitin, G. (1988) Randomness in arithmetic. *Scientific American* (July), 80–5.

Hofstadter, D. (1985) *Metamagical Themas*. Basic Books, New York.

Myhill, J. (1952) Some philosophical implications of mathematical logic. *Reviews of Metaphysics* **6**, 165–98.

■ The sound of music: hearing and listening

Ackerman, D. (1991) *A Natural History of the Senses*. Vintage, New York.

Benade, A. H. (1976) *Fundamentals of Musical Acoustics*. Oxford University Press, New York.

Moles, A. (1966) *Information Theory and Esthetic Perception*. University of Illinois, Urbana.

Parncutt, R. (1989) *Harmony: A Psychoacoustical Approach*. Springer, New York.

Pierce, J. R. (1992) *The Science of Musical Sound* (2nd edn.). Freeman, New York. (First edn. published in the Scientific American Library, 1983.)

Plomp, R. (1976) *Aspects of Tone Sensation*. Academic Press, New York.

Rigden, J. S. (1985) *Physics and the Sound of Music*. Wiley, New York.

Rossing, T. D. (1989) *The Science of Sound* (2nd edn.). Addison Wesley, Reading, Massachusetts.

Sabine, W. C. (1964) *Collected Papers*. Dover Books, New York.

Schafer, R. M. (1969) *The New Soundscape*. BMI Canada Ltd, Don Mills, Ontario.

Sundberg, J. (1987) *The Science of the Singing Voice*. Northwestern Illinois University Press, DeKalb, Illinois.

▓ Adventures of Roderick Random: white noise, pink noise, and black noise

Bolognesi, T. (1983) Automatic composition: experiments with self-similar music. *Computer Music Journal* **7**(1), 25–36.

Boon, J. P. and Decroly, O. (1995) Dynamical systems theory for music dynamics. *Chaos* **5**, 501–8.

Boon, J. P., Noullez, A. and Mommen, C. (1990) Complex dynamics and musical structures, *Interface: Journal of New Music Research*, **19**, 3–14.

Campbell, P. (1986) The music of digital computers. *Nature* **324**, 523–8.

Dodge, C., and Bahn, C. R. (1986) Musical fractals, *Byte*, June 185–196.

Gardner, M. (1978) White and brown music, fractal curves and one-over-f fluctuations. *Scientific American* **238**, (April), 16–32.

Hiller, L. A., and Isaacson, L. M., (1959) *Experimental Music*. McGraw-Hill, New York.

Hsü, K. J., and Hsü, A.J. (1990) Fractal geometry of music. *Proceedings of the National Academy of Sciences of the USA* **87**, 938–41.

Hsü, K. J., and Hsü, A. J. (1991) Self-similarity of the '$1/f$ noise' called music. *Proceedings of the National Academy of Sciences of the USA* **88**, 3507–9.

Klimontovich, Y. L., and Boon, J. P. (1987) Natural flicker noise ('$1/f$') in music. *Europhys. Lett.* **3**, 395

Knuth, D. (1984) The complexity of songs. *Communications ACM* **27**, 344–6.

Montroll, E. W., and Shlesinger, M. F. On $1/f$ noise and other distributions with long tails. *Proceedings of the National Academy of Sciences of the USA* **79**, 3380–3.

Nettheim, N. On the spectral analysis of melody. *Interface: Journal of New Music Research*, **21**, 135–48 (http://users.bigpond,net.au/nettheim/specmel/specmel.htm)

Schroeder, M. R. (1987) Is there such a thing as fractal music? *Nature* **325**, 765–6.

Schroeder, M. R. (1991) *Fractals, Chaos, Power Laws*. Freeman, New York.

Schockley, W. (1957) *Proceedings of the Institution of Radio Engineers* **45**, 279–90.

Voss, R. V. (1979) $1/f$ (flicker) noise: a brief review. In *Proceedings of the 33rd Annual Symposium on Frequency Control*, pp. 40–6. Atlantic City.

Voss, R. V., and Clarke, J. (1975) $1/f$ noise in music and speech. *Nature* **258**, 317–18.

Voss, R. V., and Clarke, J. (1978) $1/f$ noise in music: music from $1/f$ noise. *Journal of the Acoustical Society of America* **63**, 258–63.

Xenakis, I. (1971) *Formalized Music*. Indiana University Press, Bloomington.

Chapter 6: All's well that ends well

Arnheim, R. (1988) Universals in the arts. *Journal of Social and Biological Structures* **11**, 60–5.

Bak, P. (1991) Self-organising criticality. *Scientific American* (Jan.), 46–53.

Bak, P., Tang, C., and Wiesenfeld, K. (1988) Self-organised criticality. *Physical Review* **A38**, 364–4.

Berlyne, D. E. (1971) *Aesthetics and Psychobiology*. Appleton-Century-Croft, New York.

Brown, D. E. (1991) *Human Universals*. Temple University Press, Philadelphia.

Casti, J. (1994) *Complexification: Explaining a Paradoxical World through the Science of Surprise*. Harper Collins, New York.

Ehrenzweig, A. (1967) *The Hidden Order of Art*. University of California Press, Berkeley.

Gleick, J. (1987) *Chaos: Making a New Science*. Viking, New York.

Levy, S. (1992) *Artificial Life: The Quest for a New Creation*. Jonathan Cape, London.

Maquet, J. (1986) *The Aesthetic Experience: An Anthropologist Looks at the Visual Arts*. Yale University Press, New Haven.

Morse, P. (1965) *Man's Rage for Chaos*. Chilton, Philadelphia.

Tooby, J., and Cosmides, L. (1990) On the universality of human nature and the uniqueness of the individual. *Journal of Personality* **58**, 17–67.

Watts, H. A. (1980) *Chance: A Perspective on Dada*. University of Michigan Research Press, Ann Arbor.

Illustration acknowledgements

Figures

2.2	By permission of the Syndics of Cambridge University Library.
2.3	© Succession Picasso/DACS 2005.
2.5	© Tate, London 2004; © Bridget Riley 2005, all rights reserved.
2.7	*Boats at Berck-sur-Mer.* Oil on canvas, 1873, 34 × 55.8 cm. Edouard Manet, French, 1832–83. © The Cleveland Museum of Art. Purchase from the J. H. Wade Fund, 1940.534.
2.8	The National Portrait Gallery, London.
2.9	From J. D. Barrow *The Constants of Nature*, Jonathan Cape (2002) p. 123, based on information in figure 8.1 of P. D. Ward and D. Brownlee, *Rare Earth*, Copernicus (2000), p. 165.
3.3	© Getty Images.
3.4	(i) Institut für Leichtbau Entwerfen und Konstruieren, Universität Stuttgart. (i-v) From S. Hildebrandt and A. Tromba, *Mathematics and Optimal Form*, Scientific American Library; © 1985 by Scientific American Books Inc.
3.5	(i) From Julius H. Comroe (1966) The lung, *Scientific American*, **214** (February), 56–8; with permission of Jeanette W. Comroe. (ii) From Manfred Schroeder (1991) *Fractals, Chaos, Power Laws*, W. H. Freeman. (iii) © Jacky Phillips.
3.6–3.10	From G. Birkhoff, *Aesthetic Measure*, Harvard University Press (1933), with modifications.
3.11	By permission of the Syndics of Cambridge University Library.
3.12	Adapted from P. Went (1968) The size of Man, *American Scientist*, **56**, 406.
3.13	Based on data from R. May (1978) The dynamics and diversity of insect faunas, in *Diversity of Insect Faunas*, Symposia of the Royal Entomological Society of London No. 9. (ed. L. Mound and N. Waloff), pp. 188–204, Blackwell Scientific Publications.

4.11	From A. Unsöld (1977) *The New Cosmos* (2nd edn.), p. 6, Springer.
4.12	Michael T. McDermott.
4.13	From Anthony F. Aveni *Skywatchers: A Revised and Updated Version of Skywatchers of Ancient Mexico*; © 2001 Courtesy of University of Texas Press. Illustrations by Peter Durham.
4.14	From J. Laskar and P. Robutel (1993) The chaotic obliquity of the planets, *Nature*, **361**, 611; © 1993 Macmillan Magazines Ltd.
4.15, 4.16	Adapted from J. Laskar, article in *Pour la Science*, August 1993, p. 16.
4.18	Tom Lynham, 'New Sky at Night', *The Observer*, 17.7.94.
4.19, 4.20	Based on A. Roy (1984) *Vistas in Astronomy*, **27**, 172–3, figs 1 and 2.
4.21	From A. Roy (1984) *Vistas in Astronomy*, **27**, 174, figs 4 and 5.
4.22	From M. Ovenden (1966) The origin of the constellations, *The Philosophical Journal*, **3(1)** (January), 7; with permission of the Royal Philosophical Society of Glasgow.
4.23	Royal Astronomical Society.
4.24	Adapted from A. Roy (1984) *Vistas in Astronomy*, **27**, 184, fig. 9.
4.25	Adapted from R. N. Shepard (1992) In *The Adapted Mind* (ed. J. H. Barkow, L. Cosmides, and J. Tooby), fig. 13.2, Oxford University Press.
4.26	Data taken by W. N. McFarland and F. W. Munz in the summer of 1970 at Einewetok Atol. Redrawn from J. N. Lythgoe (1979) *The Ecology of Vision*, p. 5, fig. 1.2, Oxford University Press.
4.27	Based on J. N. Lythgoe (1979) *The Ecology of Vision*, p. 82, fig. 3.1, Oxford University Press; and from G. Wyszecki and W. S. Stiles, *Colour Science*, Wiley, New York.
4.29	From R. N. Shepard (1992) In *The Adapted Mind* (ed. J. H. Barkow, L. Cosmides, and J. Tooby), fig. 13.1(b), Oxford University Press.
5.1	Redrawn from Thomas A. Sebeok (1981) *The Play of Musement*, p. 225, Indiana University Press.
5.2	(a) From J. A. Hawkins (1835) *General History of the Science and Practice of Music*, Vol. 1; Dover, 1963.
5.3	Alien Musical Score, © 1992 R. Mueller. From Clifford Pickover (1992) *Mazes for the Mind*, p. 196, St Martin's Press, New York.
5.4	Based on H. C. Longuet-Higgins (1978) The perception of music, *Interdisciplinary Science Reviews*, **3**, 148–56.
5.5, 5.7	Adapted from R. Murray Schafer (1969) *The New Soundscape*, pp. 26–7, BMI Canada Ltd.
5.8, 5.9	From J. Pierce (1983) *The Science of Musical Sound*, p. 152; © 1983 by Scientific American Books.
5.10, 5.11	Adapted from Manfred Schroeder, *Fractals, Chaos, Power Laws*, pp. 111, 123, 218; © 1991 by W. H. Freeman and Company.
5.12, 5.13, 5.14	From R. F. Voss and J. Clarke (1978) $1/f$ noise in music: music from $1/f$ noise, *Journal of the Acoustical Society of America*, **63**, 259–63.
5.15	Based on N. Nettheim, *Interface: Journal of New Music Research*, **21**, 135–148 (1992).

Tables

3.1 Data from J. O. Farlow (1993) *American Journal of Science*, **293A**, 179; and from R. M. Nowak and J. L. Paradiso (1983) *Walker's Mammals of the World* (4th edn.), Johns Hopkins University Press, Baltimore.

4.1 Information from *The Planets*, Open University Press (1994), Milton Keynes.

4.2 Information from F. H. Colson (1926) *This Week*, Cambridge University Press; and E. Zerubavel (1985) *The Seven Day Circle: The History and Meaning of the Week*, Free Press, New York.

4.4 Information from B. Berlin and P. Kay (1969) *Basic Colour Terms*, p. 3, University of California Press, Berkeley.

Plates

1 Scala, Florence/HIP/The British Museum.

2 Brancacci Chapel, Florence/www.bridgeman.co.uk

3 Palazzo Ducale, Urbino/Scala, Florence, Courtesy of the Ministero Beni e att. Culturali.

4 Art Institute of Chicago/www.bridgeman.co.uk

5 The National Gallery, London/www.bridgeman.co.uk

6 © Anglo-Australian Observatory, photography by David Malin.

7 © Getty Images.

8 © ImageState.

9 Christie's Images, London/www.bridgeman.co.uk

10 © Getty Images.

11 Richard Voss.

12 © Getty Images.

13 © E. K. Thompson/Aquila Wildlife Images.

14 © A. Cardwell/Aquila Wildlife Images.

15 © M. Gilroy/Aquila Wildlife Images.

16 © Getty Images.

17 Przemyslaw Prusinkiewicz.

18 *Map C6C2 Hemisphae Alis Coeli Sphaerigra*, by permission of the British Library.

19 *Map Col 2096 Haemisphaemium Australe*, by permission of the British Library.

20 Ryoichiro Debuchi, High Tech Lab. Japan Inc. 1991.

21 From Karl von Frisch, *Animal Architecture*, Harcourt, Brace, Jovanovich, New York and London, 1974, pl 99. Photo © Dr Max Renner, Munich.

22   (a) National Gallery of Australia, Canberra. © ARS, NY and DACS, London
        2005.
23   (a) Courtesy of R. Taylor.

Every effort has been made to trace the copyright holders of unadapted material, but if any have been inadvertently overlooked the publishers will be pleased to make the necessary arrangement at the first opportunity.

Index